Artificial Intelligence in Healthcare and COVID-19

Intelligent Data-Centric Systems

Artificial Intelligence in Healthcare and COVID-19

Edited by

Parag Chatterjee

*Department of Biological Engineering, University of the Republic
(Universidad de la República), Paysandú, Uruguay*

Massimo Esposito

*Institute for High Performance Computing and Networking (ICAR),
National Research Council (CNR), Naples, Italy*

ISBN: 978-0-323-90531-2

For Information on all Academic Press publications
visit our website at https://www.elsevier.com/books-and-journals

Publisher: Mara E. Conner
Editorial Project Manager: Emily Thomson
Production Project Manager: Erragounta Saibabu Rao
Cover Designer: Miles Hitchen

Typeset by MPS Limited, Chennai, India

Working together
to grow libraries in
developing countries

www.elsevier.com • www.bookaid.org

Contents

List of contributors

Shereen Adbulla
Computer Science Department, Ploy Tech University, Erbil, Iraq

Suzan Anwar
Computer Science Department, Philander Smith College, Little Rock, AR, United States; Computer Science Department, Salahaddin University, Erbil, Iraq

Rossana Buongiorno
Institute of Information Science and Technologies "A. Faedo", National Research Council of Italy, Pisa, Italy

Rosario Catelli
Institute for High Performance Computing and Networking (ICAR), National Research Council (CNR), Naples, Italy

Sara Colantonio
Institute of Information Science and Technologies "A. Faedo", National Research Council of Italy, Pisa, Italy

Leonardo Colligiani
Department of Translational Research, Academic Radiology, University of Pisa, Pisa, Italy

Carmela Comito
National Research Council of Italy (CNR), Institute for High Performance Computing and Networking (ICAR), Rende, Italy

Srijan Das
Medical College and Hospital, Kolkata, West Bengal, India

Girirajasekhar Dornadula
Department of Pharmacy Practice, Annamacharya College of Pharmacy, Rajampeta, Andhra Pradesh, India

Massimo Esposito
Institute for High Performance Computing and Networking (ICAR), National Research Council (CNR), Naples, Italy

Salvatore Claudio Fanni
Department of Translational Research, Academic Radiology, University of Pisa, Pisa, Italy

Danila Germanese
Institute of Information Science and Technologies "A. Faedo", National Research Council of Italy, Pisa, Italy

Lakshmi Narasimha Gunturu
Scientimed Solutions Private Limited, Mumbai, Maharashtra, India

Rajani Sudhir Kamath
CSIBER, Kolhapur, Maharastra, India

Sreekantha Desai Karanam
NITTE (Deemed to be University), NMAM Institute of Technology (NMAMIT), Nitte, Karnataka, India

Dmitriy Klyushin
Taras Shevchenko National University of Kyiv, Ukraine, Kyiv

Raja Vittal Rao Kulkarni
CSIBER, Kolhapur, Maharastra, India

Silicia Lomax
University of Pennsylvania, Philadelphia, PA, United States

Mariofanna Milanova
Computer Science Department, University of Arkansas, Little Rock, AR, United States

Saja Ataallah Muhammed
Computer Science and Information Technology Department, Salahaddin University, Erbil, Iraq

Raghavendra Naveen Nimbagal
Department of Pharmaceutics, Sri Adichunchanagiri College of Pharmacy, Adichunchanagiri University, B.G. Nagar, Mandya, Karnataka, India

Andrea Pota
UOC Nefrologia e Dialisi Ospedale del Mare, Napoli, Italy

Marco Pota
Institute for High Performance Computing and Networking (ICAR)—National Research Council of Italy (CNR), Napoli, Italy

Ayla Gerk Rangel
Universidad Católica Argentina, Buenos Aires, Argentina

Chiara Romei
2nd Radiology Unit, Pisa University Hospital, Pisa, Italy

Preface

Technological evolution in our days has experienced a strong acceleration in all fields of human knowledge. In particular, the pervasiveness of artificial intelligence (AI)−based technologies has been impressive and has seen its importance grow especially in areas such as healthcare, which has had to deal with events such as the COVID-19 pandemic. During such a difficult time, the power of synergy between seemingly far-flung disciplines has made it possible to face such a calamity with vigor. But the systematic nature of an interdisciplinary approach is still far from being applied ubiquitously, and therefore, the purpose of this book is precisely to enable the dissemination of some of the methodologies underlying the research and development problems addressed recently, so as to make more widespread a method, that of a multidisciplinary approach, which becomes increasingly important to replicate to meet the challenges of the near future.

Clearly, today's health informatics is a transdisciplinary domain. With the advent of the COVID-19 pandemic, the healthcare sector across the globe has received strong shudders, and AI played an important role in dealing with the emergent challenges. Predictive data analysis, especially during the pandemic times, holds immense potential in different areas like early detection of infections, fast contact tracing, decision-making models, risk profiling of cohorts, and remote treatment. Especially in the context of the applications of AI in healthcare during COVID-19, it encounters challenges like interoperability, lack of unified structure for eHealth, data privacy, and security, making the discussion domain even wider.

In this regard, a number of relevant examples are given in this book that can be framed in three macro-areas: the first in which efforts were made to improve the predictive capabilities of the tools in use with regard to the spread and development of the different pandemic phases, the second in which the available technologies were used to develop drugs and provide assistance, and finally, the third in which the enormous amount of data available had to be properly exploited while respecting privacy on the one hand and ensuring social monitoring aid in view of the decisions to be made on the other. From a technical point of view, these three areas have relied on different AI tools applied to the areas of predictive algorithms, computer vision, and natural language processing: it is evident how the use of AI has become pervasive and fundamental to scientific advancement in any field.

In more detail, the book consists of the following nine extremely interesting chapters.

In Chapter 1 the use of software applications as a tool for surveys and data collection, as well as therapy through health tracking using AI-based techniques, has been illustrated, considering it has been extremely important to safeguard frontline medical professionals, who were even more affected by illnesses such as depression and in view of rising suicide rates during the COVID-19 pandemic.

Chapter 2 addresses an early relevant problem that urgently needed to be resolved—the prediction of epidemic curves: while traditional algorithms might have previously been sufficient, later limitations due to uncertainty about the parameters to be leveraged became a pressing problem to be solved and one that could be addressed using machine learning from real data.

Furthermore, in Chapter 3, the need to compete with time and run faster to save as many lives as possible has resulted in the need for a more incisive and rapid progression for the discovery of new drugs for use. Again, the role of AI-based technologies has made it possible to greatly improve the ability to recognize active and useful molecules for pandemic management.

In Chapter 4 the possibility of having better predictive algorithms available through machine learning has extended not only to the scenario of pandemic curves but also to that of the spread of the virus in different countries depending on precise contingent situations, which are specific according to the geography and demographics of the places, so as to help public health authorities in the work of planning and making regulations to be respected to contain the spread of COVID-19.

In Chapter 5 the use of deep neural networks has proven to be critical in providing remote telehealth in a scenario where preserving the lives of medical personnel has become as essential a constraint as being able to provide care to sick people or children unable to remotely provide a self-assessment of their health condition.

Moreover, in Chapter 6, in a scenario such as the one experienced of high social tension, the question has been raised as to how to employ the enormous amount of data conveyed through social media to monitor the overall mental and health status of the entire population, and to provide crucial support to medical and governmental institutions already hard-pressed by the overall context.

In addition, in Chapter 7, the need to make all available data effectively accessible for the study had to come to terms with the right to respect patients' privacy, and with this in mind, the use of natural language processing techniques based on deep neural networks made it possible to speed up the task of identifying all those personal data that had to be removed from medical records before making the information publicly visible.

In Chapter 8, it is shown how medical prevention, drugs and vaccines, in an effort to reduce deaths, had the dual goals of slowing down the infection and reducing its fatality, but to be effective in these purposes, it was necessary to have relevant knowledge and being able to rely on advanced and intelligent statistical methods resulting in a rather reliable future prediction regarding the evolution of the pandemic over time.

Finally, Chapter 9 illustrates how a respiratory syndrome such as COVID-19 greatly impacted the work of radiologists who needed to quickly provide a large number of diagnoses on the status of the inflammatory process in the lungs: for this reason the use of deep learning algorithms, such as image segmentation algorithms, greatly aided in the task of delineating pathology, otherwise slowed by the need for manual visual inspection that would have taken up enormous amounts of time.

The main goal of this book is to show both how collaboration in different areas of knowledge becomes essential as the specialization of fields increases in delivering sophisticated solutions, and the importance of the results that can be achieved in sensitive scenarios such as health a fortiori as threatened by a global pandemic. Undoubtedly, the development of AI has greatly expanded technical capabilities in the various fields in which it is applied, relieving humans of enormous time losses due to processes that were previously manual and are now instead carried out not only automatically but extremely quickly and punctually. It is evident that several aspects of the incorporation of AI tools and algorithms in the day-to-day healthcare paradigms, although were invoked as the need of the hour during the times of pandemic, have already initiated a paradigm shift in several areas of healthcare.

As the book is dedicated to a wide circle of readers in the most diverse fields of AI and medicine, at the same time it is intended to let everyone reflect both on the need for global cooperation and on how much there is still to be done in those fields that are more reluctant to interdisciplinary approaches and that turn out to resist technological evolution or at least fail to embrace its use as an important avenue for development. Different AI techniques, like machine learning, deep learning, and advanced areas, like computer vision and natural language processing, are extremely cross-cutting disciplines, and nowadays, more than ever their application is needed everywhere to achieve a leap forward in the quality of the tools available for us in different fields of knowledge.

The incorporation of AI in healthcare during the times of COVID-19 not only illustrated the power of AI but has also propelled a paradigm shift in modern healthcare for the years to come.

Parag Chatterjee
Massimo Esposito

Improvement of mental health of frontline healthcare workers during COVID-19 pandemic using artificial intelligence

Srijan Das[1], Silicia Lomax[2] and Ayla Gerk Rangel[3]
[1]*Medical College and Hospital, Kolkata, West Bengal, India*
[2]*University of Pennsylvania, Philadelphia, PA, United States*
[3]*Universidad Católica Argentina, Buenos Aires, Argentina*

Other notes

The background project of this work is a winner of the *MIT COVID-19 Hackathon 2020 Turning the Tide* and is currently in pilot within prominent healthcare institutions across India.

1.1 Introduction

The COVID-19 pandemic has had a profound impact on the lives of people around the world since its inception. One of the most important and negative elements that have revealed itself in the wake of the pandemic is the impact on mental health. Data from even before the pandemic have suggested that medical professionals have suffered from higher levels of psychological and psychiatric issues as compared to the general population. These problems have further grown to alarming levels during the pandemic [1,2].

There is a wide range of factors that can impact a person's mental health including experiences, which are unique to different professionals and socioeconomic backgrounds. However, due to the sheer lack of mental health professionals, it has not been possible to provide specific profession-tailored support to the people that need it, especially in developing nations. In order to address this issue, a customized support system is essential and necessary as a result of two major reasons: the first being the unique demands in addition to an existing spectrum of mental healthcare issues prevalent across a particular profession, and the second point, which is a major target in developing countries, is to have a

Artificial Intelligence in Healthcare and COVID-19. DOI: https://doi.org/10.1016/B978-0-323-90531-2.00004-7

minimum number of mental health professionals devoted to the demands of a particular team of frontline workers to support them in times of any untoward mental health emergency and also for having a constant nurturing psychological support in the long run.

Having a customized mental healthcare support system is a requisite for every professional spectrum, but more so in the case of frontline medical professionals. Several studies have demonstrated definite psychological impacts on frontline healthcare workers in terms of increased risk of suffering trauma, depression as well as anxiety [1]. A major population of such healthcare workers have reported significant psychiatric issues including suicide ideations throughout the pandemic [2]. Data regarding the rise of physician suicide rates directly linked to COVID-19 have been grossly inadequate and unorganized to date. However, the increment in levels of such cases throughout the pandemic has been palpable at large. The increase in emotional stress, personal blame for the inability to save a patient's life, and lack of feeling of self-control, in addition to many other stressors, have led to surging levels of physician burnout throughout the pandemic [3]. The physical effects of the disease COVID-19 have gained much attention; however, its effect on mental health has never gained the desired attention. The additional mental pressure is on the doctors who choose to remain away from family for months and not risk the life of the family members. They are not able to discuss and vent their worries and stress to their closest family members and are not comfortable to discuss with their colleagues, seniors, or subordinates as well.

Medical professionals, especially in the Indian subcontinent, faced some unique challenges like the stigma associated to mental illness that often leads to discrimination. Physicians' reluctance to seek mental health support is a major issue prevalent to date. Such reluctance finds roots in various preexisting stigmas and fears of potential discrimination in the workplace including consequences in obtaining a medical license [4]. Unwillingness to share problems with a colleague or a work partner is also another major problem faced by mental health professionals themselves. The rising demands of mental health assistance are also accompanied by a dearth of proper manpower to meet the same. In developing countries such as India, resident doctors of psychiatry and allied departments, belonging to major government healthcare setups, are drafted to provide medical care as well as assistance to COVID-19-specific cases. This is a major burden to countries that do not meet the ideal patient to mental health professional ratio. The final challenge faced by conventional psychiatric practices in respect to frontline workers is the inability to provide a long-term support system to the beneficiaries. Mental health issues often require long-term monitoring, assistance, and support systems, which otherwise are grossly inadequate in most healthcare setups. The long-term monitoring is in fact a prerequisite to make a complete diagnosis regarding a mental state which may in fact determine the route of therapy or support which might be most beneficial for the patient in need. An example of such a scenario is the diagnosis of posttraumatic stress disorder. An acute stress disorder persisting for more than a month is diagnosed as a posttraumatic stress disorder [5].

Owing to all such issues, it is prudent to have a system of mental health customized only for frontline healthcare workers. The system should include a method to overcome the challenges of reluctance and shortage of professional support and also replace a one-off psychiatric counseling session with a more holistic, long-term mental health support system.

1.2 **Background**

COVID-19 has arguably been one of the greatest adversities that humankind has faced in recent times. From economy and trade to policies as well as healthcare, the pandemic has been a major setback for each and every arena, globally. Not just developing countries, but even developed economies have suffered major difficulties due to the pandemic. COVID-19 was unique in the sense that humankind had not witnessed an infectious disease surge so rapidly and persistently since the times of the Spanish flu. Healthcare systems across the world have been facing immense struggles, and frontline healthcare workers have had to work in difficult conditions, often unable to care for their own health and mental stability. The proverb "physician heal thyself" has often been overlooked, and now during the pandemic, this omission has been magnified and needs to be addressed at the earliest.

Due to the pandemic, visiting a clinic or daycare service is not possible for frontline healthcare workers due to overpacked schedules. The other hurdles that come along while approaching professional help have also been discussed above. Thus, the option of telemedicine and application-based monitoring systems needs to be explored.

Artificial intelligence can play a major role as a savior. It can help in digital discussion and can be a platform of support for the medical professionals with easy constant access. The most important advantage of using artificial intelligence (AI) is its ability to gather huge amounts of data and prediagnose the risk of developing mental distress among the individual physicians and combat the problem more efficiently [6]. To make the support more easily and readily accessible with a minimal cost, the authors have tried to make a mobile app: the Khushi: Cura te Ipsum mobile application.

Recently in 2021, two apps have been designed by AIIMS, New Delhi, Shaksham and Disha, for patients with first episode of symptoms and those having severe mental illness. However, still now no such apps have been specially designed for our frontline workers. They are the ones who deal with both patients including mental health patients, but they do not have a platform to share and support their own problems. So is our endeavor for Khushi.

Quantitative data in form of paper-based questionnaire and statistical analysis for developing mental health apps have also been done where 176 participants including 88 patients and 88 caregivers have been surveyed [7].

The advent of telemedicine in recent times has been undeniable. However, due to the rising demands and increasing patient load, it is not possible for a professional to provide adequate time to a patient in need. Also, telemedicine itself is unable to guarantee long-term monitoring, and thus, a constant state of monitoring may not be achieved. The aforementioned factors along with various other issues have led people to shift to more automated and digitized platforms such as mobile applications (apps) [8]. A major factor responsible for the increased popularity of app-based platforms, especially during the times of COVID-19 pandemic, is the fact that there is an increased willingness to embrace digital health services during disasters [9]. The role of a smartphone application and its advantages over a desktop-based system are numerous. Given that the approximate population of smartphone users is 6 billion as compared to 2 billion personal computers worldwide, the former is definitely more accessible. As the pandemic took a more vicious shape, the demand for mental health applications surged. The rise in popularity of apps such as Calm-Meditate, Sleep, Relax has been phenomenal such that it achieved the second-highest worldwide rank among all health and fitness apps on the Google Play Store grossing sales worth over US$ 1,149,000 [10]. The shift from a clinic-based approach to a more automated and accessible one has thus been successful and promising.

However, there exists an entirely unexplored arena with huge potential for support and profit, which is the mental healthcare of frontline medical workers. There is a lack of tailored apps targeting such demography. Mental healthcare for frontline workers has thus been restricted to the conventional methods or amalgamated with general demands of an uncategorized population, and thus, specific demands have not been taken into account. Upon this background, it has been absolutely essential to introduce a holistic mental healthcare initiative in the shape of a mobile-based application designed specifically for frontline healthcare workers which will help in overcoming various obstacles in addition to providing a long-term monitoring system.

Searching into existing literature, the authors observed a rise in the popularity of app-based platforms focused on mental health and well-being [11−14]. For instance, Calm-Meditate, Sleep, Relax was on track to hit a $150 million revenue run rate in 2019, and for the year 2020, they had estimated revenue of $150 to $200 million. Another acclaimed health app, Headspace, generated $100 million revenue in 2019 [15,16].

Additionally, as the COVID-19 pandemic took a more vicious shape, the demand for mental health applications has increased [9]. Taking into consideration the previous applications mentioned as examples, the app Calm-Meditate, Sleep, Relax achieved the second-highest worldwide rank among all health and fitness apps on the Google Play Store, causing its sales to be approximately worth over US$ 1,149,000 in 2019 [9]. This application, in 2019, has been valued at 1 billion and in the year 2020 at $2 billion. In 2019, 40 million users downloaded the app, and in 2020, 100 million downloaded it. In 2019, the app had 2 million subscribers, and in 2020, it doubled its numbers to 4 million subscribers.

Headspace also had an increase in its demand. In the year of 2018, Headspace had 1 million subscribers and 2 million subscribers in 2020. Concerning downloads, they registered 40 million in 2018 and 65 million in 2020 [15,16]. This exponential increase in these healthcare technological features could be explained due to the fact that there is an increased willingness to embrace digital health services during disasters [8].

There are various applications available today which cater to mental health. Additionally, another aspect that has been considered is the surge in demand for such services throughout the year 2020 [17,18]. Furthermore, as a result of research into the literature and after profound analysis of preexisting software, applications, and existing needs, the authors got too many enlightening conclusions. This generated an idea and plan to achieve the goals of reducing stress and other mental health challenges although this time focusing on frontline healthcare workers.

1.3 Main content

The creation of the app Khushi: Cura te Ipsum is to help and support the long-neglected mental distress among the medical professionals on whose hands our life depends. The need for such apps became more important during the pandemic.

The point considered during the creation of the Khushi: Cura te Ipsum app is the rise of mental health applications regarding the possible dissemination of misinformation that could harm the users. The authors have thereby identified the importance of properly trained mental healthcare professionals and clinicians to get more involved and serve as subject matter experts and thereby develop the app further.

One of the many positive aspects of the creation of an app such as Khushi: Cura te Ipsum would be to integrate and use telehealth via an application-based platform to deliver mental healthcare for health professionals. This will help to surpass and solve many barriers still present in telehealth. This application would initially focus on low- and middle-income areas, which the authors consider to be the places with higher necessity and the ones to benefit the most even with existing barriers. Research has proven that even with all the restrictions, low- and middle-income countries have benefited from technology in the medical field, especially in COVID-19, in which, for instance, telehealth has shown to be a valuable solution for many of the existing obstacles presented. If we take into consideration the nations with larger territories, an acute deficit of healthcare professionals, together with an inequitable distribution, could be seen as a difficulty to implement such technology. Nevertheless, reality has proven the opposite. Telehealth, technology, and AI along with many other technological features have shown positive results, in remote and low-income areas, in many aspects such as increasing compliance of patients, improvement of medical decisions as well as a reduction of healthcare costs, which is necessary since one of the biggest concerns prevails in the financial aspects of healthcare [19,20].

Keeping the aforementioned prerequisites and looking into previous literature to understand which features could be more beneficial by looking into the usage, patterns, and retention metrics of frequently installed mental health apps, the authors propose a mobile-based application to monitor the mental health of frontline healthcare workers in the long run. This would have a multidimensional approach in the sense that it includes both preventive and therapeutic approaches [21].

The aspects other than the demography and the multidimensional outlook which make the app unique in its approach are its endeavor to address the problems of reluctance to approach a mental health professional and that ignoring initial warning signs of an impending mental health disorder by the use of continuous survey-based systems powered by AI.

Regarding AI, results have shown that AI is an important digital solution to approach mental health prediction, detection as well as treatment. Moreover, it helps in enhancing and optimizing personalized experience [22–25]. Another advantage of AI is that it helps in surpassing the barriers of professional shortage with this technological help. AI has also been shown to facilitate financial, geographical, and temporal limitations, additionally being an aid to the unfortunate existent stigmatization barriers. However, it is important to understand its boundaries, such as the concerns regarding confidentiality, and furthermore acknowledge that AI is not a replacement for mental healthcare professionals.

Further evidence found in existing literature concerns the benefits of combining components in platforms, also known as hybrid solutions, which offered a blend of face-to-face and online delivery as an effective solution [26,27]. Moreover, a distinct action that has shown to be useful in improving patient outcomes is measuring mood over the long run [28,29]. This has been shown to have increased awareness and knowledge about the user's own mental health and provide multiple benefits in terms of the therapy. Nevertheless, concerns about mood tracking require the need for preparation from beforehand. More information and instructions are needed in advance so that users can understand, prepare themselves, and gain more insights about their personal mental health. Other concerns pertain to the lack of orientation for action points, in which experts advise the need for guidelines on how to manage crises. Furthermore, the sustainability of user engagement and consistency along with insufficient personalization consists of other possible disadvantages of mood trackers.

1.4 Methodologies and implementation

The creators have been considering a distinct solution to approach mental health prediction and provide the best outcome possible. The collection and interpretation of metadata in addition to surveys (active data) would help the application achieve so. Consequently, it would provide knowledge of how the patients complete the survey and thereby provide more clinical specificity and information. Moreover, metadata has been shown to be advantageous due to the generation of high volume of automated data and fewer ethical challenges (as compared to passive data), and it is a feature that still has not been explored extensively, thereby giving this app another distinguished benefit as compared to most of the existent applications in the market [13].

For the reasons mentioned previously in detail, it may be concluded that a hybrid application, which offers a combination of face-to-face interaction, online healthcare delivery, alongside AI and motivator features with engaging resources, will generate a unique application as shown in Fig. 1.1.

This multidimensional project will include a total of three components that address the issues of regular monitoring: reluctance and hurdles in approaching a mental health professional by providing an initial platform of beginning a conversation and finally by providing holistic and proven mind-calming preventive and supportive measures.

The first methodology is regular monitoring which would be performed by mental health surveys sent personally to the app on a weekly basis. The survey form will be received by every healthcare provider and must be solved within a preset period. The survey would be designed to be a simple, compact yet effective tool for screening initial levels of psychological stress, anxiety, and depression. For this purpose, the authors propose to use the DASS-21 scale, (with modifications suitable for the region and workplace) as a single instrument to measure the emotional states of depression, anxiety, and stress.

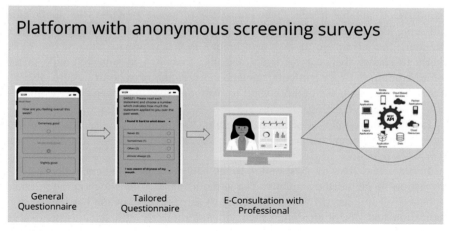

FIGURE 1.1

Sequence of collecting information and working of the application.

These weekly mental health surveys will collect the necessary information on a regular basis which would be then assessed by the virtues of AI. The role of AI in this regard is to assess mood, check initial signs of anxiety, depression, and increasing stress level, and address them at the earliest. A cut-off score will be assigned for the scale, and cases with scores above the same would be detected as the ones requiring active psychological support. Once detected, the application would direct the selected cases to be assessed by a psychiatrist or mental health professional working at the hospital or connected via the platform. This helps to capture the dynamic sense of mental health and extend care beyond the clinic. The role of AI is indispensable in this regard. The initial screening by AI will not only improve the accuracy but also reduce the dependence upon the professional mental healthcare workers and in the meanwhile, bring to the forefront, various psychological issues which might not have been discovered by the patient himself/herself.

The second methodology deals with improving compliance and aiding the person in need to approach a professional. As discussed above, the authors have identified several problems which a person faces while reaching out to a mental health professional and how it is more pertinent in cases of a physician himself/herself. To overcome these barriers, the application plans to work on a second methodology which involves building an anonymous matching module between the beneficiary and the service provider. Anonymity guarantees patients to reach out to professionals with greater confidence and lesser reluctance. Further virtues of the matching module include providing more power to make a choice regarding the mental healthcare provider, something which is nearly entirely absent in conventional clinical practice. The role of AI would also be essential since only those professionals with the required skill set as chosen by the beneficiary would be recommended. After that, the user can switch through the profiles to find themselves a suitable match. Once connected to a care provider, the application will serve as an initial platform for engaging in discussions with a registered care provider. The advantage of such a technique is that it provides the patient more liberty to choose their care provider and also to switch to any preferable match before committing to a particular therapeutic regimen or support system. By following this method, the authors want to overcome any inconvenience or uneasiness in setting up a dialog.

Lastly, the presence of additional features, highlighting the use of short prerecorded mind-calming techniques, will serve as therapeutic support throughout their work period.

In order to deliver the best option to the patients and customers, more features are also to be included eventually. The authors consider their incorporation could be beneficial in demystifying and helping to surpass some of the barriers of stigmatization and discrimination, which are unfortunately still an intrinsic reality and issue in many societies, especially some lower- or middle-income countries. Furthermore, extra features could make the application more attractive, appellative, and helpful in the engagement and commitment to the app. The concern relies on the loss of its purpose and singularity as using extra features could approximate many other applications already existent in the market. Regarding this aspect, the developers looked into previous literature to understand which features could be more beneficial by looking into the usage, patterns, and retention metrics of frequently installed mental health apps [21]. Hence, the final method is a holistic approach in both therapeutic and preventive sense. This would primarily be based on mind-calming techniques such as guided meditation and yoga which have been proven to be effective in preventing depression, anxiety, and stress [30]. The recorded messages would be accessible to the patients which can be used as per requirements. This would also be supplemented by an anonymous forum where people can confide and vent their emotions.

Another service that the application developers plan to include is the emergency button. This emergency button deals with any sort of mental health emergencies that the user may face. On using the button, an emergency response system would be activated which would connect the person in need directly to an attending psychiatrist on a high-priority basis. It will help in preventing unnecessary delays in the dispatch of service and prevent any untoward events from occurring.

Alongside all these services, the app will also provide users with an index of useful contacts which belong to helplines dealing with suicide prevention and organizations dealing with deaddiction.

Concerning the technical aspects of this application, one of the targets is that even if it is created for iPhone or Android, it would also be effective for Windows or Mac. In this case, the authors exposed the backend as a REST API which means that the app has a single backend and that the application may be used on all platforms, therefore dispensing the need to create separate entities, resulting in saving development time, and, furthermore, improving efficiency and portability. This area is in the process of improvement and progress, with many ideas and possibilities to increase the application to its full potential in order to deliver the best possible results.

The application primarily seeks to provide two types of subscription in form of individual memberships and institution-based memberships. The target audience is the healthcare workers belonging to the institutions who have subscribed for the services of the app and the healthcare workers who are individually subscribed to the application. To ascertain their profession as frontline healthcare workers, identification would be collected during the process of registration. Once a specific count of subscribers is available, the application would employ the ideal workforce required to meet the demands of the former. The authors want to propose a sustainable and self-dependent system based on the integration of the psychiatric and psychology clinic of the same establishment. However, the authors identify the absence of a proper mental workforce in hospitals in developing countries and the absence of mental health departments in various setups, especially the ones catering to a singular departmental service other than mental healthcare. The authors recognize the importance of recognizing and understanding this aspect and how it would help in planning the medium dispatch of service. Hence, they looked at India's ratio of patients to psychiatrists to have a broader comprehension of the demands of a developing nation. Evidence shows that India has 0.75 psychiatrists per thousand people, while the desirable number is anything above three psychiatrists per thousand. This is extremely low once compared to high-income countries where the estimated numbers found are around six psychiatrists per 100,000 people. Considering that the evidence suggests the desirable number of psychiatrists would be at least three psychiatrists per 100,000 population, 36,000 would be approximately the number of psychiatrists required to reach the ideal goal. Data show that India is currently short of 27,000 doctors based on the current population of the country. This is a concern and could be considered a barrier to implementation. The literature, as previously mentioned, suggests technology as a potential solution to this obstacle [31,32]. Moreover, the thought of having a dedicated workforce of hired health professionals would be implemented as an additional strategy.

The precedence of the professional is still discussed, as having a psychiatrist from the same hospital or not having in consideration what would be more beneficial for the patients. The possibility of professionals from the same hospital as the patients raises concerns of confidentiality, stigmatization and personal comfort. In order to generate the best possible outcome, it is necessary to reflect on these aspects, such as, if the patient would feel comfortable to ask for help even knowing that who would be in charge of their treatment works in the same place as they do. Nevertheless, on the other hand, it would be beneficial considering a logistic point of view and plan implementation perspective, facilitating many aspects of the strategy chain. In conclusion, it was decided that in cases of hospitals, clinics, and institutions with an adequate workforce, the application aims to achieve a sustainable self-dependent model, whereby the mental health professionals will be able to cater to their own setup but on the grounds of maintaining anonymity through the platform. In such a case, the application will simply serve as an anonymous medium for frontline workers to get the required help.

In cases of setups with inadequate or absence of mental healthcare staff, the application plans to serve as a medium for connecting with a professional who would be contractually employed under the app.

Keeping in mind that some institutions have dedicated care to psychological and psychiatric aspects, the authors would want to enter into a partnership with the same. However, the ideal ratio of psychiatrists to patients is something that the application would always seek to achieve. The application will employ under its purview mental health professionals who cater to the demands of beneficiaries. Thus the business model which the app follows is primarily based on transactions between the institution and the developers or a singular subscriber and the developer (Fig. 1.2).

Regarding financing and average revenue profit, hospital administrative staff would be incentivized to maintain their staffing levels and minimize turnover by adequately providing services for them to continue high-quality work within their

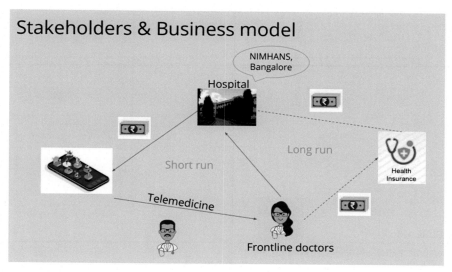

FIGURE 1.2

Business model of the application.

facility. This business model provides a payment platform that contains a bulk membership option for all of the frontline workers to be included and easily access the same through their mobile devices. After an introductory period of approximately 6 months to 1 year, the frontline workers and administrative staff would have the opportunity to provide the application developers feedback on its positive impact and how it can be improved in the future. If the application proves to be satisfactory, then the hospital facility has the option to continue the relationship with the application but can move to a long-term model that is included through the employee health insurance as a part of the premium payment from the employees so that it is simultaneously paid for through the employer and employee and provides benefits to both entities. This model allows for sustainable usage of the application by the frontline workers as well as regular payments to the application to continue its developments and enhancements.

One of the main steps during the implementation of the application was found to be the development of a prototype. It would be undertaken in a preselected institution to observe the integration, technical aspects, barriers, benefits, as well as necessary improvements. For developers to understand the application before launching, the possibility of a free feature opportunity in which they could test and offer feedback, such as a user feel, would be considered. Once the application is ready and organized, institutional ethical clearance would be filed for proceeding with the same.

The implementation strategy (Fig. 1.3) for successfully launching the application is clear-cut and made by keeping in mind the economic problems faced by

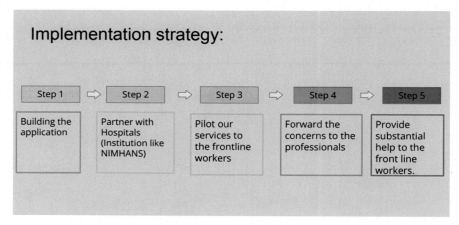

FIGURE 1.3

Implementation strategy of the application.

developing economies and their healthcare. The implementation would take place in five steps. The first and foremost step would be to complete building the application by taking into account the needs for modification suggested by the experts in the field. The second step would be to partner with government hospitals or institutions, preferably ones that are dedicated to mental healthcare and have a suitable population of mental health professionals to function with. This step involves the ethical clearance on behalf of the institutional ethics committee. The third step would be to pilot the project in a controlled setup taking into account a specific number of beneficiaries and professionals involved under the purview of the app. The fourth step would involve employing and creating a professional workforce under the app. This would be done in two routes. The former is the integration of the mental healthcare workforce of the same institution, and the latter is contractually employing professionals. The final step involves the introduction of the app among the beneficiaries so that substantial mental support could be provided to the same. The last step involves promotion and letting frontline healthcare workers know about the benefits of the application. The application would be subjected to feedback and recommendations from the users and the employees. Data would be collected on a regular basis and analyzed by the team. Suitable modifications can then be included to improve the quality of service.

The developers forecast valuable results not only from a business perspective, but primarily as the main concern and aim, on delivering the best possible outcomes for the patients targeted.

Serving as one of the core functions of this application, Khushi: Cura te Ipsum has regular monitoring of the user's mental health status through a survey. This hypothetical graphical (Fig. 1.4) depiction of the mood of an arbitrary user named Sunita illustrates that over time, the feelings of stress and anxiety

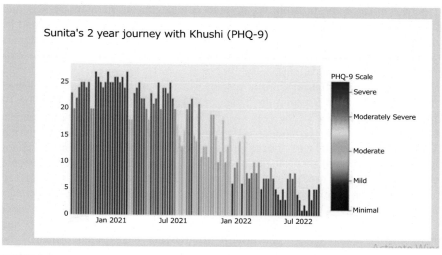

FIGURE 1.4

An auto-generated patient profile window correlating with usage of the application.

significantly declined from the combined features of the application and lifestyle changes. This is meant to be shown to a combination of the user, the mental health professional, and the application developers. Fig. 1.4 shows there were severe negative symptoms from January to July 2021 and then improvements in 2022.

A limitation of this article, besides the present obstacles this application faces, such as workforce availability, confidentiality, development, and social stigmatization along with other barriers, consists of the lack of usable statistics about previous consultations. Hence, a prototype will be needed to collect such information and data. Nevertheless, the pivot study has already been planned to be undertaken in government hospitals and medical colleges after attaining proper ethical clearance on behalf of the ethical clearance committee. The authors have already approached state healthcare authorities for starting a pivot study in return for free service for a negotiable period. The authors are currently awaiting permission from an ethical and administrative viewpoint, to begin with, the implementation. Furthermore, additional existing planning projects of future expansion to other countries are yet to be planned upon. The authors foresee that by providing this unique multidisciplinary approach outstanding results could be achieved. Khushi: Cura te Ipsum application would thus be able to help many healthcare workers to manage their mental health struggles in this unprecedented scenario but also at any time.

Additionally, it is necessary to highlight as the central matter of this app focuses on the importance of patient confidentiality and how to protect it. One of the strategies assumed to maintain this statement consists of a starting point in the download of the app, which will be anonymous. However, certain issues would be persistent once the possibilities of proceeding with an anonymous IRB and trailing hospitals, feedback systems are considered as well as the use of AI and its difficulties regarding the protection of confidentiality, although the developers will consistently work toward this main aspect in order to deliver the best outcomes to the users and patients.

The application, Khushi: Cura te Ipsum, was a brainchild of finalists selected for the event of India: Turning the Tide, a Hackathon under the MIT COVID-19 Challenge. The premise of the app was decided by considering the requirements of India and similar economies into consideration. Problems faced during COVID-19 were brought to notice, and solutions were proposed from a technological, economic, and medical perspective by a team which consisted of people belonging to diverse professional backgrounds. The ideas were further polished and developed into deliverables with the help of international mentors volunteering under the MIT COVID-19 Challenge. Team Khushi was declared the winner of India: Turning the Tide, MIT COVID-19 Challenge, among participants and finalists from 60 countries. The criteria for judging were based primarily on impact, innovation, implementation, and presentation. After being declared winners, the developers were offered mentorships by various professionals who wish to contribute toward the building of the application. The journey of *Khushi: Cura te Ipsum* was also featured by media outlets which provided a further boost to the future of the application. The future of the application, thus, seems promising with various enthusiastic partnership options, and thus, the authors believe that the idea has what it takes to be a groundbreaking solution to the ongoing crisis of mental health.

1.5 Discussion

The results from preliminary findings of the COVID app indicate that there are promising solutions that can come from the use of the application and the users that are impacted by it in both their interpersonal relationships and their professional lives. The mental health of our essential workers and healthcare professionals is necessary to monitor and try to keep on the positive end of the spectrum as it has the greatest impact on the population. The necessity of this application is eminent and gives us reason to want to maintain the best possible outcomes for this group of individuals that are targeted. As there are a number of possibilities that can come with the use of this application, the strengths and weaknesses must be considered as well.

1.5.1 Connection to artificial intelligence

There are several elements of AI that are incorporated into this application. It is meant to be a fully customizable experience to give the user the feeling and understanding that it is designed not only to improve mental health generally, but also to tailor the design to an individual's experiences and perspectives of their daily interactions [33]. The use of the surveys in the application is to gather that initial measure of the person's challenges in life, their beliefs, and some of their personal interests. This serves multiple purposes for the user. The first is to identify where on the depression, stress, and anxiety scale do they fall in order to get them the attention and professional assistance that they need. The second purpose is to then lead to user-specific features in the application based on the beliefs and interests that they share. The more of this information that is shared, the more that the technology can tailor it to their experiences [25]. For example, a user can indicate that one of the methods for them to feel calm is if they listen to music. The application can then sync the user up to a music experience, but it may be generic in nature. If a user specified that they like music of a specific genre or artist, then the application can link them to that genre or artist for a better experience. This is the same case for meditation services, games, and other features that may assist the user in these other ways. This use of AI also helps as the application will constantly learn from the elements of the application that are frequently activated by users and make those more accessible or share push notifications that allow them to engage with it on a more regular basis [34]. The application can also deactivate certain features on the application if there is no use after a limited number of days. It is not a feature that is permanently removed but is meant to evolve with the person that is using it.

1.5.2 Strengths

The most notable strength of this application is the usage of AI to make those more customized experiences for users and lead to better interaction with it [25]. The other strength of the application is that it is a more regular measure of someone's mental health measures. Rather than only one baseline measure of their mental health, there are weekly measures of how someone is doing on the spectrum. People's feelings of anxiety, depression, and stress levels are impacted by changes that happen in their lives as well as in anticipation of certain events. In order to address these constantly varying feelings, these measures must be taken on a regular basis to keep up with the demand of their bodies and overall health.

Another strength is the number of different features within the application. It is meant to become a tool that can be used by anyone and has a number of features to attract people with different interests. It is also an incredibly portable platform in which getting access to the care that you need can be more immediate and operational on a device that practically every adult carries with them. This can be accessed via iPhone or Android and has the opportunity to be tested on other devices as well with its expansion. The emotion tracking feature is great to monitor by both the user and on the backend as a way to constantly improve. The music, meditation, and other activities within the application are significant for daily use especially as COVID-19 has exacerbated feelings of loneliness and isolation. This application gives people the opportunity to feel more connected as well as short activities that will not take too much valuable time from the days of these valuable workers.

1.5.3 Weaknesses

While there are a number of remarkable strengths with this application to address mental health in the healthcare workforce, there are also weaknesses that must be considered. One of the most impactful concerns of this application is confidentiality and privacy. The user would have to share a significant amount of personal information in order for the assessment of their mental health to be accurate. Additional information is also needed for the AI within the application to make it a customizable experience for the user. This could lead to privacy concerns by the user since this information will have to be tracked in order to share their overall changes and improvements in mental health. Although the developers of the application can promise not to share this information externally, the fact that the information is going into any database may still be disconcerting.

Although this is outside of the capabilities of the developers of this application to fully address, another weakness is the limited number of mental health professionals. There is a very finite number of resources in the current mental health workforce, and their capacity may be met before all of the frontline healthcare professionals impacted by COVID-19 can receive the attention that they need. The application team could attempt to remedy this challenge by offering shorter interactions with mental health professionals as needed in order for more people to be seen. There is also potential in administering a volunteer base of mental health trainees to establish a larger cohort of individuals that can assist the frontline workers that are seeking that care.

The application is currently designed to cater to a specific population that only includes frontline healthcare workers. This is primarily to positively impact the challenges that they have had to face in the wake of the COVID-19 pandemic. As the piloting of the application takes place and improvements are made, there is a possibility that the application can expand beyond this population to people within other professions. This may especially be of use in high-stress environments that require constant quick-response thinking and the necessity to remain calm regardless of circumstance. This expansion may be explored and can be used to inform the initial functions of the application so that the AI is applied to anyone in other working functions.

1.6 Conclusion

Since its inception, the COVID-19 pandemic has had devastating effects on the world, causing a massive economic decline, an overutilization of healthcare systems, and detrimental challenges to people's mental health. This held to be especially true for frontline healthcare workers that tirelessly aimed to disrupt the rampage of the COVID-19 pandemic on the health of the world's citizens. Many of their lives were lost, and they were forced to make immeasurable sacrifices in order to care for the people that contracted the highly infectious disease leading to harmful effects on their own health. The recognition of these challenges is what led to the thought leadership and beginning stages of the development of the Khushi: Cura te Ipsum mobile application.

The purpose of the mobile application, *Khushi: Cura te Ipsum*, is to provide a solution to the emerging mental health and limited access issues among the frontline healthcare workers all throughout the pandemic. There is one primary reason that the application is unique when compared to other applications that are currently available. It is because it focuses on a specific job demographic, which is the frontline healthcare workers only. Frontline healthcare workers were chosen because the authors believe that the health of physicians, nurses, and other medical professionals is often ignored and consistently overlooked as a legitimate issue. Based on previous studies, describing the burnout rates of healthcare professionals prior to the pandemic and the evidence that surfaced about the challenges brought on by the pandemic illustrate the dire need to have this application for the frontline healthcare workers now. This is specifically due to the fact that there has been a surge in mental health issues among medical health professionals during the pandemic [35].

The solution is unique in the sense that it is multidimensional in its approach. As described in the methodology and the discussion above, the application carries multiple features across the platform that provide self-servicing usage for improved mental health as well as the connection to a mental health professional or trainee that can provide additional assistance based on the responses in the regularly planned surveys. It also achieves the required goal without increasing the burden on the already strained and understaffed mental healthcare departments in various nations, especially in the developing economies. It does so by relying on the virtues of AI. The initial screening and warning signs are detected by AI on a regular basis and then succeeded by a final assessment and evaluation of the case by a mental health professional. This reduces a direct dependence on the latter and also detects cases which otherwise would not have been considered serious due to the lack of a regular monitoring system. This dynamic approach not only helps to cater to a singular case with more accuracy, but also helps to keep in account the general mental satisfaction scale of the entire workplace alongside the dynamic preventive approach. There are several methods that the app follows to have a holistic approach to all the problems that may otherwise be uncatered to in common psychological and psychiatric practices.

The system ensures a minimum availability of mental healthcare professionals by assigning a team to a singular setup of frontline workers. Instead of achieving a huge target of ensuring a proper ratio of professionals to patients at once, the authors sought to divide the problem into a smaller demographic for a specific career so as to cater to them individually. The application would primarily be a service tailored to a singular healthcare setup that has subscribed for the service.

With the use of AI, a specific level of sorting would be performed depending upon the requirements and priorities set by the user surrounding their answers to the survey questions on their mental health and their daily lives. Following this sorting stage, the application will also serve as a platform for establishing an initial anonymous conversation between the beneficiary and the mental health professional. The conversation can vary in length based on the needs of the beneficiary and serves as the stage to set the baseline for what the frontline worker may need to definitively improve. This will then be followed by a conventional psychiatric evaluation and recommendations to enhance the lifestyle and the usage of features in the application to contribute to that as well.

There are still elements of the application that can be further developed to cater to this population and ensure an improvement in the overall mental health of frontline workers. As the piloting phases continue and there is feedback to improve the usage of the application, it will be able to fully service the people that it is meant to assist. Although the design is specifically for this demographic, the lessons from the enhancements made to the application may one day be applied to other demographics and for other particularly high-intensity professions that need this regular monitoring and access to features. One thing that is certain from other studies and lived experiences is that even if the pandemic is fully over, the application will still be an incredibly useful resource to this demographic. The challenges with mental health and burnout rates happened long before the start of the pandemic and will likely continue afterward [36]. This application uses technology and AI to offer that glimmer of hope to populations that have desperately needed its access to improve their mental health challenges and lead to better life outcomes.

References

[1] S. Cabarkapa, S.E. Nadjidai, J. Murgier, C.H. Ng, The psychological impact of COVID-19 and other viral epidemics on frontline healthcare workers and ways to address it: a rapid systematic review, Brain Behav. Immun. (2020) 100144.

[2] K.P. Young, D.L. Kolcz, D.M. O'Sullivan, J. Ferrand, J. Fried, K. Robinson, Health care workers' mental health and quality of life during COVID-19: results from a mid-pandemic, National Survey, Psychiatr. Serv. 72 (2) (2021) 122−128.

[3] M.A. Reger, I.H. Stanley, T.E. Joiner, Suicide mortality and coronavirus disease 2019—a perfect storm? JAMA Psychiatry 77 (11) (2020) 1093−1094.

[4] S.S. Mehta, M.L. Edwards, Suffering in silence: mental health stigma and physicians' licensing fears, Am. J. Psychiatry Residents J. 13 (11) (2018) 2−4.

[5] F. Edition, Diagnostic and statistical manual of mental disorders, Am. Psychiatr. Assoc. (2013) 21.

[6] A.B.R. Shatte, D.M. Hutchinson, S.J. Teague, Machine learning in mental health: a scoping review of methods and applications, Psychol. Med. (2019) 1−23. Available from: https://doi.org/10.1017/S0033291719000151.

[7] K. Sinha Deb, A. Tuli, M. Sood, R. Chadda, R. Verma, S. Kumar, et al., Is India ready for mental health apps (MHApps)? A quantitative-qualitative exploration of caregivers' perspective on smartphone-based solutions for managing severe mental illnesses in low resource settings, PLoS One 13 (9) (2018) e0203353. Available from: https://doi.org/10.1371/journal.pone.0203353. PMID: 30231056; PMCID: PMC6145572.

[8] T. Basu, The coronavirus pandemic is a game changer for mental health care, Hum. Technol. (2020) 20.

[9] J. Torous, M. Keshavan, COVID-19, mobile health and serious mental illness, Schizophrenia Res. (2020).

[10] Statista Leading Health and Fitness Apps in the Google Play Store Worldwide in March (2020). By revenue: Statista (2020). Available online at: https://www.statista.com/statistics/695697/top-android-health-apps-in-google-play-by-revenue/ (accessed April 11, 2020).

[11] K.K. Weisel, L.M. Fuhrmann, M. Berking, H. Baumeister, P. Cuijpers, D.D. Ebert, Standalone smartphone apps for mental health—a systematic review and meta-analysis, NPJ Digit. Med. 2 (1) (2019) 118.

[12] M. Milne-Ives, C. Lam, M.H. Van Velthoven, E. Meinert, Mobile apps for health behavior change: protocol for a systematic review, JMIR Res. Protoc. 9 (1) (2020) e16931.

[13] J. Torous, H. Wisniewski, B. Bird, E. Carpenter, G. David, E. Elejalde, et al., Creating a digital health smartphone app and digital phenotyping platform for mental health and diverse healthcare needs: an interdisciplinary and collaborative approach, J. Technol. Behav. Sci. 4 (2) (2019) 73−85.

[14] Center for Devices, Radiological Health. Device software functions and mobile medical applications [Internet]. Fda.gov. 2020 [cited 2021 Apr 22]. Available from: https://www.fda.gov/medical-devices/digital-health-center-excellence/device-software-functions-including-mobile-medical-applications.

[15] Calm revenue and usage statistics (2021) [Internet]. Businessofapps.com. 2020 [cited 2021 Apr 23]. Available from: https://www.businessofapps.com/data/calm-statistics/.

[16] Headspace revenue and usage statistics (2021) [Internet]. Businessofapps.com. 2020 [cited 2021 Apr 23]. Available from: https://www.businessofapps.com/data/headspace-statistics/.

[17] Apa.org. [cited 2021 Apr 23]. Available from: https://www.apa.org/monitor/2021/01/trends-mental-health-apps.

[18] L.C. Ming, N. Untong, N.A. Aliudin, N. Osili, N. Kifli, C.S. Tan, et al., Mobile health apps on COVID-19 launched in the early days of the pandemic: content analysis and review, JMIR MHealth UHealth 8 (9) (2020) e19796.

[19] C.P. Chandrasekhar, J. Ghosh, Information and communication technologies and health in low income countries: the potential and the constraints, Bull. World Health Organ. 79 (9) (2001) 850−855.

[20] C.O. Bagayoko, D. Traoré, L. Thevoz, S. Diabaté, D. Pecoul, M. Niang, et al., Medical and economic benefits of telehealth in low- and middle-income countries: results of a study in four district hospitals in Mali, BMC Health Serv. Res. 14 (Suppl 1(S1)) (2014) S9.

[21] A. Baumel, F. Muench, S. Edan, J.M. Kane, Objective user engagement with mental health apps: systematic search and panel-based usage analysis, J. Med. Internet Res. 21 (9) (2019) e14567.

[22] S. D'Alfonso, AI in mental health, Curr. Opin. Psychol. 36 (2020) 112–117.

[23] C.A. Lovejoy, V. Buch, M. Maruthappu, Technology and mental health: the role of artificial intelligence, Eur. Psychiatry 55 (2019) 1–3.

[24] A. Gamble, Artificial intelligence and mobile apps for mental healthcare: a social informatics perspective, Aslib J. Inf. Manag. 72 (4) (2020) 509–523.

[25] S. Graham, C. Depp, E.E. Lee, et al., Artificial intelligence for mental health and mental illnesses: an overview, Curr. Psychiatry Rep. 21 (2019) 116.

[26] J. Torous, K. Jän Myrick, N. Rauseo-Ricupero, J. Firth, Digital mental health and COVID-19: using technology today to accelerate the curve on access and quality tomorrow, JMIR Ment. Health 7 (3) (2020) e18848.

[27] D.C. Mohr, K.R. Weingardt, M. Reddy, S.M. Schueller, Three problems with current digital mental health research... And three things we can do about them, Psychiatr. Serv. 68 (5) (2017) 427–429.

[28] N. Rickard, H.-A. Arjmand, D. Bakker, E. Seabrook, Development of a mobile phone app to support self-monitoring of emotional well-being: a mental health digital innovation, JMIR Ment. Health 3 (4) (2016) e49.

[29] C. Caldeira, Y. Chen, L. Chan, V. Pham, Y. Chen, K. Zheng, Mobile apps for mood tracking: an analysis of features and user reviews, AMIA Annu. Symp. Proc. 2017 (2017) 495–504.

[30] V. Lemay, J. Hoolahan, A. Buchanan, Impact of a yoga and meditation intervention on students' stress and anxiety levels, Am. J. Pharm. Educ. 83 (5) (2019).

[31] K. Garg, C.N. Kumar, P.S. Chandra, Number of psychiatrists in India: baby steps forward, but a long way to go, Indian J. Psychiatry 61 (1) (2019) 104–105.

[32] M.A. Hoffer-Hawlik, A.E. Moran, D. Burka, P. Kaur, J. Cai, T.R. Frieden, et al., Leveraging telemedicine for chronic disease management in low- and middle-income countries during Covid-19, Glob. Heart 15 (1) (2020) 63.

[33] D.D. Luxton, An introduction to artificial intelligence in behavioral and mental health care, Artificial intelligence in behavioral and mental health care, Academic Press, 2016, pp. 1–26.

[34] S. D'Alfonso, AI in mental health, Curr. Opin. Psychol. (2020).

[35] N. Greenberg, M. Docherty, S. Gnanapragasam, S. Wessely, Managing mental health challenges faced by healthcare workers during covid-19 pandemic, BMJ. (2020) 368.

[36] S. De Hert, Burnout in healthcare workers: prevalence, impact and preventative strategies, Local Reg. Anesthesia 13 (2020) 171.

Effective algorithms for solving statistical problems posed by COVID-19 pandemic

2

Dmitriy Klyushin

Taras Shevchenko National University of Kyiv, Ukraine, Kyiv

2.1 Introduction

Apparently, COVID-19 has become not an episode in human history, but its constant companion, such as the flu. The rapid and unpredictable spread of the coronavirus puts many governments in a difficult position, forcing them to make decisions that threaten the economic well-being of people. As a result, the significance of accurate forecasting of the COVID-19 outbreak has increased significantly. Standard compartmental models (SIR, SEIR, etc.) contain parameters that depend on poorly predictable factors—the degree of population mobility, sanitary measures taken, etc. Obviously, these parameters can be estimated only very approximately. In such situations, there is increasing attention to alternative approaches, in particular to machine learning models. These models do not make assumptions about the shape of the epidemic curve and uncertain parameters, using only observable data.

Every forecasting model generates a time series and accuracy measures (the standard error [MSE], the mean absolute percentage error [MAPE], and the mean absolute deviation [MAD]). Then, they are compared to each other when choosing a model having higher accuracy. However, when comparing time series, it is necessary to analyze both the integral accuracy measures and the distribution of errors. If the samples of model errors are statistically equivalent, then it is meaningless to compare them in accuracy. On the other side, if the errors have different distributions, then the selection of the most accurate predicting model is statistically justified.

In this chapter, we develop effective nonparametric two- and k-sample homogeneity tests and use them to estimate the forecasting accuracy of compartment forecasting models and machine learning forecasting models for the prediction the epidemic curves in several countries.

Section 2.1 is introduction. Section 2.2 is devoted to the main problems in forecasting epidemic curves. Subsection 2.2.1 contains an overview of papers on

Artificial Intelligence in Healthcare and COVID-19. DOI: https://doi.org/10.1016/B978-0-323-90531-2.00005-9

21

forecasting models used for the prediction of COVID-19 epidemic curve. In subsection 2.2, we describe nonparametric tests for comparison of forecasting models. Section 2.3 outlines nonparametric tests used for forecasting model estimation and versions of the Klyushin—Petunin test of sample homogeneity. In Section 2.4, we apply the versions of the Klyushin—Petunin test, the Kolmogorov—Smirnov test, and the sign test for estimation of the time series accuracy and arrange models for predicting the coronavirus epidemic curves in some countries for a given period. Section 2.5 contains conclusions and description of future work.

2.2 Forecasting the epidemic curves of coronavirus

The key problems in predicting the coronavirus epidemic curve are the complexity and validness of predictive models [1]. The validness and robustness of the predictive model for the coronavirus epidemic curve have great significance since governments make important decisions using these prognoses. Thus, correct estimation of the prediction quality is an urgent problem.

2.2.1 Forecasting models for the COVID-19 outbreak

The most known models for predicting infectious diseases are the SIR and SEIR compartment models and their derivatives [2]. The letters S, E, I, and R mean that a model takes into account susceptible (S), infected (I), recovered (R), and exposed (E) compartments. The SIR [3] and SEIR [4] models are systems of differential equations depending on the transition rates between compartments.

The SIR and SEIR standard models and their derivatives have been widely used to predict COVID-19 outbreaks in many countries [5—11]. They are mathematically strong and precise, but they depend on uncertain factors. The epidemic curves predicted by the compartment models are mathematically sound, but they often are significantly differing from true epidemic curves. Now, the most widely used epidemiological models are the SIR, SEIR models, and their derivatives. They are standard models in this area [12]. The drawbacks of the standard models follow from the high parameters' uncertainty. That is why epidemiologists frequently do not use predicted time series directly and prefer to simulate "best and worst" scenarios. In the opposite to systems of differential equations, the machine learning algorithms use only real training data giving more precise results.

Consider some recent achievements in forecasting the COVID-19 pandemic using the compartment models. Li et al. [5] inferred epidemiological characteristics of SARS-CoV-2 using observations of reported infection cases in China, mobility data, fraction of undocumented infections, the contagiousness of the infection, a networked dynamic metapopulation model, and Bayesian inference.

Anand et al. [6] developed a new version of the SIR model taking into account the proportion of infected individuals in India. In opposite to the classical SIR model, their approach ignores not a realistic assumption of homogenous mixing of population and makes forecasts more precise. Ifghuis et al. [7] applied the SIR model in the Kingdom of Morocco and proved that the conventional SIR model much more accurately predicts the final size of the COVID-19 outbreak than the SEIR model. As we see, the key factor of forecasting precision is the reality of the assumptions of the SIR model. If these assumptions are not satisfied, the SIR model becomes rather a theoretical tool.

There were some attempts to improve the compartment models using a statistical approach. Nesteruk [8] identified the optimal values of the SIR model parameters taking into account the numbers of infected, susceptible, and removed persons and has made a short-time prediction of the development of coronavirus epidemic in China using exact solution of the SIR equations. As noted in Ref. [8], the exactness of the prediction models cannot be guaranteed if their parameters are based on non-reliable data. Babu et al. [9] carried out a short-term prediction of new incident cases of COVID-19 for India using the logistic growth curve model prediction and the SIR model. The authors state that the logistic growth curve model was more accurate than the SIR model because the SIR model over-estimated the number of new incident cases. He et al. [10] developed the modified SEIR model taking into account both epidemical and environmental factors and showed that it made adequate forecasts for some regions of China, Italy, South Korea, and Iran. Also, the authors stressed the vital importance of correct data and noted that the data in many cases have large uncertainty. Guirao [11] made an attempt to predict the trend of the COVID-19 epidemic and to estimate the risk of the outbreak. He used the SIR or SEIR models and proposed two analytical formulas that allow easy estimation of the final size of the epidemic in possible scenarios and provide a theoretical background for the decision-making. Wang et al. [12] reviewed modifications of the SIR and SEIR mathematical models for the COVID-19 outbreak. The authors have discovered that the mathematical modeling for the COVID-19 outbreak is still a difficult problem, and it is based mostly on epidemic dynamic models rather than statistical ones or machine learning. Like most other investigators, the authors have marked the considerable uncertainty of estimation of the epidemiological characteristics in the models for the COVID-19 outbreak.

Many investigators believe that machine learning models are more effective if forecasting the COVID-19 pandemic because they do not use unrealistic assumptions. For example, Swapnarekha et al. [13] analyzed numerous statistical, machine learning, and deep learning forecasting models and described their applications for the prediction of the COVID-19 pandemic. The authors state that machine learning, deep learning, and statistical approaches are very effective for the prediction of the COVID-19 pandemic. They prove that fact by citing numerous sources noting, meanwhile, the scarcity of the epidemical data. Sujath et al. [14] presented an original model for the prediction of the spread of SARS-CoV-2.

The authors have implemented linear regression (LR) model, multilayer perceptron (MLP), and vector autoregression method for the COVID-19 Kaggle data to forecast the rate of COVID-19 cases in India. The MLP has provided more precise predictions than the LR model and vector autoregression methods. Thus, the authors suppose that using the deep learning methods for forecasting time series data allows us to obtain more precise estimations of the COVID-19 pace.

Ardabili et al. [15] compared several outbreak prediction models for COVID-19 that received the most attention from authorities and media (SIR and SEIR). The authors have noted a high level of uncertainty and, therefore, low precision of the conventional models for long-term prediction. They presented a comparative analysis of machine learning and SIR and SEIR models to predict the COVID-19 epidemic. The authors choose MLP and adaptive network-based fuzzy inference systems as the most prospective methods. In the opinion of Ardabili et al., due to highly complex and variable nature of the SARS-CoV-2 machine learning methods are more effective tool to forecast the pandemic than the compartmental models. The paper contains the benchmark data to demonstrate the prevalence of the machine learning methods. Tuli et al. [16] developed an improved mathematical model to predict the rate of the COVID-19 epidemic and applied it to data gathered in some countries. The authors have deployed a cloud computing platform for their mathematical model and carried out accurate and real-time prediction of the epidemic rate. They have shown that a data-driven approach provides high accuracy and can be very useful for decision-making. Distante et al. [17] carried out modeling of the COVID-19 outbreak in different Italian regions. They trained a deep convolution autoencoder with COVID-19 data in China and used the SEIR model to predict the spreading and peaks. Training neural networks on data from China and the knowledge about the spreading of COVID-19 in Italia have provided a good fit. This paper is a rare example where compartment models and machine learning models were not opposed but used together.

Artificial neural networks are widely used for prediction of the COVID-19 outbreaks. Kolozsvári et al. [18] predicted the epidemic curves of the COVID-19 pandemic using recurrent neural networks (RNNs) and validated the predicted models with the observed data. The authors show that the errors between the predicted and validated data are low. The authors note that their machine learning model more accurately predicts the peak of the epidemic curve than conventional model. Ibrahim et al. [19] introduced an original variational LSTM Autoencoder model to predict the spread of the coronavirus that uses not only relying on historical data but also takes into account population factors (density, fertility, etc.) and governmental decisions (lockdown, restrictions, etc.). The proposed models have shown high accuracy in short- and long-term forecasting and can be useful for decision-making. Hu et al. [20] developed artificial intelligence methods for modeling the transmission dynamics of the COVID-19 pandemic. The developed methods were applied to the data on the COVID-19 cases and deaths reported by the WHO in March 2020. The authors state that real-time forecasting by their methods is more accurate than standard epidemiologic models, and such models can be used in policymaking concerning the COVID-19 pandemic. Guo et al. [21] developed an artificial neural network for modeling the confirmed infection cases of COVID-19 and deaths from COVID-19. This study shows that the ANN model has suitable precision for predicting the COVID-19 outbreak because predictions are very close to the actual data.

Among other papers devoted to the application of machine learning methods to the COVID-19 outbreak prediction, we would mark the work of Balli [22] where a time-series forecasting model using machine learning is developed. Balli used LR, MLP, random forest, and support vector machine (SVM) learning methods to obtain the epidemical curve of COVID-19. According to the standard accuracy metrics, SVM achieved the best results. The main goal of Kafieh et al. [23] was to forecast the outbreak in Iran, Germany, Italy, Japan, Korea, Switzerland, Spain, China, and the United States using MLP, random forest, and different versions of long short-term memory (LSTM). The authors compared the performances of the models using standard accuracy measures, including mean average percentage error (MAPE), root mean square error (RMSE), normalized RMSE (NRMSE), and R^2. The best precision was provided by a modified version of LSTM.

Rustam et al. [24] demonstrated the capability of machine learning models to predict COVID-19 incident cases. The authors compared LR, least absolute shrinkage and selection operator (LASSO), SVM, and exponential smoothing (ES) to forecast the COVID-19 outbreak. The results prove that these methods are very effective tools for predicting and analyzing possible scenarios of the COVID-19 pandemic.

As we see, forecasting the coronavirus pandemic requires new effective tools for model evaluation. These methods should estimate error values and their distribution without additional assumptions. Nonparametric statistics tests satisfy such requirements. Nonparametric tests are very useful in econometrics [25,26], climate forecasting [27], and other scientific disciplines to assess prediction errors. These tests do not make any assumptions on error distributions and preconditions, except most general ones, for example, unimodality of a random value.

To estimate forecasting models, we propose effective algorithms based on Hills's assumption $A_{(n)}$. To justify our point of view, we applied these algorithms to forecasts provided by the intelligent optimization method gray wolf optimizer (GWO) and two machine learning models of forecasting time series: MLP and adaptive network fuzzy inference system (ANFIS) are considered [15]. The GWO is an optimization algorithm that simulates the hunting pack of wolves [28]. Among machine learning models, Ardabili et al. chose a MLP and an ANFIS [29]. For our purposes, it does not matter which algorithms were used to predict the coronavirus pandemic, as we will be comparing the homogeneity of their errors.

2.3 Nonparametric tests used for forecasting models estimation

When forecasting time series, we must choose the most accurate model. Since the precision of the model by nature is a random variable, we can compare two models with different distributions of accuracy only by testing the hypothesis, that is, their homogeneity. For identical error distributions, models should be considered equivalent even if they exhibit different average accuracy. Traditional measures of model accuracy are MSE, MAPE, and MAD. Thus, a comparative analysis of the accuracy of the models can be reduced to testing the hypothesis about the homogeneity of samples containing forecast errors. This is assumed that the model errors are stationary and unbiased.

Diebold and Mariano [25] proposed the following procedure for testing the equality of forecasting models accuracy. Let $T_j, j = 1, ..., m$ be forecasting models, $t_i^{(j)}$ be predictions of a data sequence t_i, $i = 1, ..., n$ produced by the model T_j and $\varepsilon_i^{(j)}$, $i = 1, ..., n; j = 1, ..., m$ be errors of the model T_j obeying a distribution F_j. Introduce a loss function $J(\varepsilon_i^j)$ depending on the model accuracy. The main assumption about identical accuracy of models T_k and T_l is $r_i^{(k,l)} = J\left(\varepsilon_i^{(k)}\right) - J\left(\varepsilon_i^{(l)}\right) = 0$. Therefore, $E\left(J\left(\varepsilon_i^{(k)}\right)\right) = E\left(J\left(\varepsilon_i^{(l)}\right)\right)$. When as a loss function we use the standard deviation $\varepsilon_i^{(k)} - \varepsilon_i^{(l)}$, we reduce the problem to the testing of the hypothesis $E\left(\varepsilon_i^{(k)}\right) = E\left(\varepsilon_i^{(l)}\right)$. However, the assumption on identical forecasting accuracy may be generalized and reduced to test the general null hypothesis $F_k = F_l$.

2.3.1 Nonparametric tests for homogeneity

There are numerous tests for testing the homogeneity of two samples that are subdivided into several groups: permutation tests, rank tests, randomization tests, and distance tests. Also, the tests may be consistent with any pair of alternatives (such as Kolmogorov−Smirnov test [30,31]), and with pairs of different alternatives of a given class (Dixon [32], Wald and Wolfowitz [33], Mathisen [34], Wilcoxon [35], Mann−Whitney [36], Wilks [37], etc.). Also, they could be divided into two large groups: purely nonparametric and conditionally nonparametric. Nonparametric tests do not use any assumption on distribution functions [30−37]. Conditionally, nonparametric tests (Pitman [38], Lehmann [39], Rosenblatt [40], Dwass [41], Fisz [42], Barnard [43], Birnbaum [44], Jockel [45], Allen [46], Efron and Tibshirani [47], Dufour and Farhat [48]) use some assumptions on distributions.

The two-sample Kolmogorov−Smirnov test [30,31] is purely nonparametric test. It compares empirical cumulative distribution functions constructed on two samples. Test statistics estimate the maximum difference between these functions.

The advantage of the Kolmogorov—Smirnov test is that it is based on no assumptions about the shape of the function, but at the same time, it is too sensitive to outliers. The Kolmogorov—Smirnov test works perfectly when samples have different means. Instead, the Wald—Wolfowitz test [33] is more powerful than the Kolmogorov—Smirnov test when samples have almost the same location and different variances and vice versa.

As we shall see, this point is a key feature for the estimation of the performance of a test. The group of tests for location shift contains the Mathisen test [34], the Wilcoxon signed-rank test [35], the Mann—Whitney test [36], the Wilks test [37], etc. When the assumption on the same location is violated, it is reasonable to use other tests that correctly take into account this fact.

Many publications note that the Kolmogorov—Smirnov test has the maximum sensitivity if the mean values of the two samples differ greatly from each other; that is, the test statistic reaches its greatest value in a vicinity of a center of the distribution, and not in the area of its tails. If the average values are almost the same, then the Kolmogorov—Smirnov test becomes ineffective. To eliminate this deficiency, the Cramer—von Mises test [49,50] was proposed, which estimates the sum of the squared deviations between empirical distribution functions, rather than the modulus of the difference between them. However, it is also insensitive when differences between samples are mainly concentrated far from the center of the distribution, since the asymptotes of the empirical distribution functions are zero and one so that differences far from the center of the distribution are leveled. This problem is solved by the Anderson—Darling test [51,52], which is a variant of the Cramer—von Mises test with weights. Like the Kolmogorov—Smirnov test, it is nonparametric and at the same time has a higher sensitivity. At the same time, computations by the Anderson—Darling test in the case of small samples are associated with computational difficulties.

In addition to low sensitivity in cases where the test statistic takes a maximum value in the tail region, the Kolmogorov—Smirnov test has another drawback. If the distribution functions of both samples are continuous, then the test indeed does not depend on the distribution, but the limiting distribution of the test statistic is difficult to calculate. For discrete populations, the Kolmogorov—Smirnov test depends on an underlying distribution. Since the Cramer—von Mises statistics is a weighted Kolmogorov—Smirnov statistic, it has the same disadvantages.

To resolve the difficulties, Allen developed tests that use the distances between probability densities in different spaces [46]. Unlike the Kolmogorov—Smirnov and Cramer—von Mises tests, which are nonparametric only if the distribution functions of the samples are continuous, the Allen tests are nonparametric only if the samples are infinite. For finite samples, these tests are sample dependent and therefore are conditionally nonparametric. The third category of tests uses the difference between the mean sampled values (e.g., [38,41,47]). If the samples are finite, then all the tests of this group depend on distributions; that is, they are not nonparametric. In order to level this dependence, Dufour and Farhat [48] proposed to use arbitrary permutations of the elements of the combined sample. Since all these permutations have the same probability, a test rejecting the null hypothesis using a critical value obtained on a conditional distribution of given order statistics may be considered nonparametric.

As the analysis of nonparametric and conditionally nonparametric tests shows, there are several problems in this area: (1) hypothesis testing is often associated with significant computational difficulties; (2) the properties of the tests are usually asymptotic; and (3) not all tests are valid against all possible alternatives. To solve these problems, we propose the following test.

2.3.2 Exact nonparametric test for homogeneity

Hill's assumption $A_{(n)}$ [53] states that for exchangeable random values $x_1, x_2, ..., x_n$ obeying an absolutely continuous distribution, the following equation holds:

$$P(x_{n+1} \in (x_{(i)}, x_{(j)})) = p_{ij} = \frac{j-i}{n+1}, \quad j > i, \tag{2.1}$$

where x_{n+1} and $x_1, x_2, ..., x_n$ are identically distributed, and $x_{(i)}$ is the i-th order statistics. Consider samples $x = (x_1, x_2, ..., x_n)$ and $y = (y_1, y_2, ..., y_m)$ drawn from absolutely continuous distributions F_1 and F_2. Introduce an event $A_{ij}^{(k)} = \{x_{(i)} < y_k < x_{(j)}\}$ and its relative frequency $h_{ij}^{(k)}$. Constructing an exact confidence interval $I_{ij}^{(n,m)}$ for the binomial proportion of the event $A_{ij}^{(k)}$, we can compute the frequency L of an event $\left\{ p_{ij} = \frac{j-i}{n+1} \in I_{ij}^{(n,m)} \right\}$. Denote by N the number of all the possible pairs of order statistics $x_{(i)}$ and $x_{(j)}$. This number is equal to $(n-1)n/2$. Therefore, the relative frequency $\rho(x, y) = L/N$ is a homogeneity measure of samples x and y which we shall call the exact P-statistics. To obtain a symmetrical version of the exact P-statistics and balance the role of two samples, swap x and y and find $\rho^*(x, y) = \frac{1}{2}(\rho(x, y) + \rho(y, x))$.

Events $A_{ij}^{(k)}$ form the generalized Bernoulli scheme [54] if the null holds. If the null hypothesis fails, this scheme is called the modified Bernoulli scheme [55]. In the general case, when the null hypothesis can be either true or false, this scheme is called the Matveichuk–Petunin scheme [56]. Thus, the decision rule of the test for the null hypothesis $F_1 = F_2$ with a significance level $\alpha < 0.05$ has the form: if $h > 0,95$ the null hypothesis is accepted, else the null hypothesis is rejected.

There are 20 confidence intervals for the binomial proportion [57]. The confidence interval for the binomial proportion often depends on the parameter which varies from one distribution to another. For example, the parameter in the Wilson interval varies from 1.96 (for standard normal distribution) to 3.0 (for general unimodal distribution). To avoid this uncertainty, we propose to use the exact confidence interval for the unknown probability constructed using the relative frequency of success in the Bernoulli model consisting of trials [58]. This interval is determined by formulas (2.2)–(2.6).

When calculating the confidence interval for the binomial proportion, it is necessary to find a compromise between the desire to cover all values of the binomial proportion with a probability of at least a given confidence level (coverage probability) and the length of this interval. Confidence intervals that have a probability of coverage greater than 0.95 for all sample sizes and binomial values are said to be exact. The most common two-sided exact confidence interval is the Clopper–Pearson interval. Note, however, that when calculating the confidence interval, the complexity of the calculations is no less important. The Clopper–Pearson interval requires rather complex calculations of special functions. Thus, there is a need for a method for constructing accurate intervals for the binomial proportion, which would be quite simple and at the same time provide a relatively short interval length.

Consider the functions

$$\varphi(p) = |h - p|, \quad \psi(p) = \frac{1}{2n} + \frac{\lambda}{n}\sqrt{np(1-p) + \frac{1}{12}}, \text{ and } \tilde{\psi}(p) = \sqrt{np(1-p) + \frac{1}{12}}, \quad p \in R^1.$$

(2.2)

The graph of $\tilde{\psi}(p)$ is the upper half of an ellipse E passing through the points $A = \left(\frac{1}{2n}\left(n + \sqrt{\frac{n}{3} + n^2}\right), 0\right)$, $B = \left(\frac{1}{2}, \sqrt{\frac{1}{12n} + \frac{1}{4}}\right)$, $C = \left(\frac{1}{2n}\left(n - \sqrt{\frac{n}{3} + n^2}\right), 0\right)$, and $D = \left(\frac{1}{2}, -\sqrt{\frac{1}{12n} + \frac{1}{4}}\right)$ with a center $\left(\frac{1}{2}, 0\right)$. The graph $\psi(p)$ is a narrowing of the graph of $\tilde{\psi}(p)$ on the segment $[0, 1]$ with shrinking or stretching by factor $\frac{\lambda}{n}$ and shifting by value $\frac{1}{2n}$. Therefore, the graph of the function $\psi(p)$ which does not depend on h is an arc of ellipse Γ passing through the points $(0, \psi(0))$, $\left(\frac{1}{2}, \psi\left(\frac{1}{2}\right)\right)$, $(1, \psi(1))$, such that the function $\psi(p)$ reaches the minimum at the point $p = \frac{1}{2}$, and it is symmetrical with respect to this point.

The lower confidence limit p_1 of the exact interval is a root of the quadratic equation.

$$\left(1 + \frac{\lambda^2}{n}\right)p^2 - \left(\frac{\lambda^2}{n} - \frac{1}{n} + 2h\right)p + h^2 - \frac{h}{n} + \frac{1}{4n^2}\left(1 - \frac{\lambda^2}{3}\right) = 0.$$

(2.3)

If $h > \psi(0) = \frac{1}{2n} + \frac{\lambda}{n\sqrt{12}}$, then the lower confidence limit p_1 is the least root of (3). If $h \leq \psi(0)$, then $p_1 = 0$.

Similarly, the upper confidence limit p_2 of the exact interval is a root of the square equation.

$$\left(1 + \frac{\lambda^2}{n}\right)p^2 - \left(\frac{\lambda^2}{n} + \frac{1}{n} + 2h\right)p + h^2 + \frac{h}{n} + \frac{1}{4n^2}\left(1 - \frac{\lambda^2}{3}\right) = 0.$$

(2.4)

If $1 - h > \psi(1)$, then the upper confidence limit p_2 is the largest root of (4). If $1 - h \leq \psi(1)$, then $p_2 = 1$.

Remark: Note that $p_1 \leq h \leq p_2$, so that the proportion of successes always lies in the confidence interval $[p_1, p_2]$.

For the generalized Bernoulli model, similar reasoning gives the following quadratic equation for lower confidence limit:

$$\left(1 + \frac{(m+n+1)\lambda^2}{(n+2)m}\right)p^2 + \left(\frac{1}{m} - \frac{(m+n+1)\lambda^2}{(n+2)m} - 2h\right)p + h^2 - \frac{h}{m} + \frac{1}{4m^2}\left(1 - \frac{\lambda^2}{3}\right) = 0$$

(2.5)

If $h > \frac{1}{2m} + \frac{\lambda}{m\sqrt{12}} = \gamma$, then the lower confidence limit p_1 for the generalized Bernoulli model is the least root of Eq. (2.5). If $h \le \gamma$, then $p_1 = 0$.

Similarly, the upper confidence limit p_2 in the generalized Bernoulli model is the root of the equation.

$$\left(1 + \frac{(m+n+1)\lambda^2}{(n+2)m}\right)p^2 - \left(\frac{1}{m} + \frac{(m+n+1)\lambda^2}{(n+2)m} + 2h\right)p + h^2 + \frac{h}{m} + \frac{1}{4m^2}\left(1 - \frac{\lambda^2}{3}\right) = 0$$

(2.6)

If $1 - h > \gamma$, then the upper confidence limit p_2 is the largest root of Eq. (2.6). If $1 - h \le \gamma$, then $p_2 = 1$. By virtue of the previous results, the significance level of the confidence interval does not exceed $\frac{4}{9}\frac{1}{\lambda^2}$ (in particular, 0.05 for $\lambda = 3$).

The decision rule of the exact test for the null hypothesis $F_1 = F_2$ with a significance level $\alpha < 0.05$ has the form: if $\rho^*(x, y) > 0,95$ the null hypothesis is accepted, else the null hypothesis is rejected.

2.4 Comparison of forecast models

For an illustration of the performance of the proposed exact nonparametric test, consider the errors of prediction of the COVID-19 outbreak by the version of GWO, MLP, and ANFIS methods. As benchmarks for comparison, we use the Klyushin–Petunin test (the P-statistics) [59], the Kolmogorov–Smirnov test, and the sign test.

We use the training data (Ardabili et al., 2020) for China, Germany, Italy, Iran, and the United States on total cases of disease over 30 days and prediction over 150 days. For comparison, we consider the results for several time moments with the step 20 days. Therefore, we have samples consisting of eight errors. The null hypothesis on the identity of distribution functions was accepted if the P-statistics (the probability that the samples are drawn from the same population) in the Klyushin–Petunin test and the proposed exact nonparametric test is greater than 0.95, and the significance level $(1 - P)$ of the Kolmogorov–Smirnov test and sign test is less than 0.95. The results are provided in Tables 2.1–2.20. Here, LogitGWO denotes logistic GWO, LinGWO denotes linear GWO, LogGWO denotes logarithmic GWO, QGWO denotes quadratic GWO, PGWO denotes power GWO, MLP denotes multilayer perceptron, and ANFIS denotes adaptive network fuzzy inference system.

Tables 2.1–2.5 demonstrate that almost all the methods are statistically different (P-statistics is less than 0.95 in 84.7% of cases). The machine learning methods in three cases (Italy, Germany, and China) of five are statistically equivalent.

Table 2.1 The *P*-statistics for the forecast models of the COVID-19 outbreak in Italy.

Method	LogitDWO	LinGWO	LogGWO	QGWO	PGWO	MLP	ANFIS
LogitGWO	1.000	0.893	0.786	0.857	0.992	0.893	0.813
LinGWO	0.893	1.000	0.786	0.893	0.992	0.821	0.821
LogGWO	0.786	0.786	1.000	0.786	0.821	0.786	0.786
QGWO	0.857	0.893	0.786	1.000	1.000	0.786	0.786
PGWO	0.992	0.992	0.821	1.000	1.000	0.750	0.750
MLP	0.893	0.821	0.786	0.786	0.750	1.000	1.000
ANFIS	0.813	0.821	0.786	0.786	0.750	1.000	1.000

Using the P-statistics form the data on the COVID-19 outbreak in Italy, we may conclude that (1) the distributions of errors of the GWO models are different in 7 of 10 cases; (2) the distributions of errors of the machine learning models differ from the distributions of errors of all the GWO models; (3) the distributions of errors of the machine learning models do not differ from each other.

Table 2.2 The *P*-statistics for the forecast models of the COVID-19 outbreak in Germany.

Method	LogitDWO	LinGWO	LogGWO	QGWO	PGWO	MLP	ANFIS
LogitGWO	1.000	0.750	0.893	0.893	0.786	0.536	0.893
LinGWO	0.750	1.000	0.786	0.893	1.000	0.536	0.786
LogGWO	0.893	0.786	1.000	0.786	0.893	0.536	0.892
QGWO	0.893	0.893	0.786	1.000	0.856	0.489	0.786
PGWO	0.786	1.000	0.893	0.856	1.000	0.536	0.821
MLP	0.536	0.536	0.536	0.489	0.536	1.000	1.000
ANFIS	0.893	0.786	0.892	0.786	0.821	1.000	1.000

Using the P-statistics for the data on the COVID-19 outbreak in Germany, we may conclude that (1) the distributions of errors of the GWO models are different in all the cases; (2) the distributions of errors of the machine learning models differ from the distributions of errors of all the GWO models; (3) the distributions of errors of the machine learning models do not differ from each other.

Table 2.3 The *P*-statistics for the forecast models of the COVID-19 outbreak in Iran.

Method	LogitDWO	LinGWO	LogGWO	QGWO	PGWO	MLP	ANFIS
LogitGWO	1.000	0.893	0.786	0.821	0.786	0.786	0.857
LinGWO	0.893	1.000	0.786	0.929	0.893	0.821	0.821
LogGWO	0.786	0.786	1.000	0.786	0.786	0.786	0.786
QGWO	0.821	0.929	0.786	1.000	1.000	0.786	0.786
PGWO	0.786	0.893	0.786	1.000	1.000	0.750	0.750
MLP	0.786	0.821	0.786	0.786	0.750	1.000	0.893
ANFIS	0.857	0.821	0.786	0.786	0.750	0.893	1.000

Using the P-statistics for the data on the COVID-19 outbreak in Iran, we may conclude that (1) the distributions of errors of the GWO models are different in all the cases; (2) the distributions of errors of the machine learning models differ from the distributions of errors of all the GWO models; (3) the distributions of errors of the machine learning models differ from each other.

Table 2.4 The *P*-statistics for the forecast models of the COVID-19 outbreak in China.

Method	LogitDWO	LinGWO	LogGWO	QGWO	PGWO	MLP	ANFIS
LogitGWO	1.000	0.786	0.786	0.786	0.786	1.000	0.964
LinGWO	0.786	1.000	0.786	1.000	1.000	0.750	0.750
LogGWO	0.786	0.786	1.000	0.786	0.786	0.786	0.786
QGWO	0.786	1.000	0.786	1.000	1.000	0.750	0.750
PGWO	0.786	1.000	0.786	1.000	1.000	0.750	0.750
MLP	1.000	0.750	0.786	0.750	0.750	1.000	0.964
ANFIS	0.964	0.750	0.786	0.750	0.750	0.964	1.000

Using the P-statistics for the data on the COVID-19 outbreak in China, we may conclude that (1) the distributions of errors of the GWO models are different in 8 of 10 cases; (2) the distributions of errors of the machine learning models differ from the distributions of errors of almost all the GWO models, except one case; (3) the distributions of errors of the machine learning models do not differ from each other.

Table 2.5 The *P*-statistics for the forecast models of the COVID-19 outbreak in the United States.

Method	LogitDWO	LinGWO	LogGWO	QGWO	PGWO	MLP	ANFIS
LogitGWO	1.000	0.750	0.857	0.893	0.750	0.964	1.000
LinGWO	0.750	1.000	0.786	0.893	1.000	0.786	0.786
LogGWO	0.857	0.786	1.000	0.786	0.856	0.893	0.893
QGWO	0.893	0.893	0.786	1.000	0.857	0.786	0.786
PGWO	0.750	1.000	0.856	0.857	1.000	0.786	0.786
MLP	0.964	0.786	0.893	0.786	0.786	1.000	0.857
ANFIS	1.000	0.786	0.893	0.786	0.786	0.857	1.000

Using the P-statistics for the data on the COVID-19 outbreak in the United States, we may conclude that (1) the distributions of errors of the GWO models are different in 9 of 10 cases; (2) the distributions of errors of the machine learning models differ from the distributions of errors of almost all the GWO models, except one case; (3) the distributions of errors of the machine learning models differ from each other.

Table 2.6 The exact *P*-statistics for the forecast models of the COVID-19 outbreak in Italy.

Method	LogitDWO	LinGWO	LogGWO	QGWO	PGWO	MLP	ANFIS
LogitGWO	1.000	1.000	0.893	0.929	0.929	0.964	1.000
LinGWO	1.000	1.000	0.893	0.964	0.964	0.893	0.893
LogGWO	0.893	0.893	1.000	0.893	0.929	0.893	0.893
QGWO	0.929	0.964	0.893	1.000	1.000	0.893	0.893
PGWO	0.929	0.964	0.929	1.000	1.000	0.857	0.857
MLP	0.964	0.893	0.893	0.893	0.857	1.000	1.000
ANFIS	1.000	0.893	0.893	0.893	0.857	1.000	1.000

Using the exact P-statistics for the data on the COVID-19 outbreak in Italy, we may conclude that (1) the distributions of errors of the GWO models are different in 6 of 10 cases; (2) the distributions of errors of the machine learning models differ from the distributions of errors of almost all the GWO models, except one case; (3) the distributions of errors of the machine learning models do not differ from each other.

Meanwhile, they significantly differ from the GWO methods. It should be noted that the versions of GWO methods are significantly different in most cases. The fact that the distributions of model errors, mostly, are different allows for comparing these models by accuracy. As far as all conclusions in most cases are similar, the data obtained in the mentioned countries may be considered consistent.

Tables 2.6−2.10 demonstrate that most of the models are statistically different (the exact *P*-statistics is less than 0.95 in 80.9% of cases). The machine learning methods in four cases of five are statistically equivalent; meanwhile, they significantly differ from the GWO methods in most cases. Note that the versions of GWO methods are significantly different in less number of cases comparing with the Klyushin−Petunin test. Possibly, this may be explained by the fact that the exact

Table 2.7 The exact *P*-statistics for the forecast models of the COVID-19 outbreak in Germany.

Method	LogitDWO	LinGWO	LogGWO	QGWO	PGWO	MLP	ANFIS
LogitDWO	1.000	0.857	0.964	0.964	0.823	0.536	0.964
LinGWO	0.857	1.000	0.893	0.964	1.000	0.536	0.893
LogGWO	0.964	0.893	1.000	0.893	0.964	0.536	0.964
QGWO	0.964	0.964	0.893	1.000	0.964	0.500	0.893
PGWO	0.823	1.000	0.964	0.964	1.000	0.536	0.929
MLP	0.536	0.536	0.536	0.500	0.636	1.000	1.000
ANFIS	0.964	0.893	0.964	0.893	0.929	1.000	1.000

Using the exact P-statistics for the data on the COVID outbreak in Germany, we may conclude that (1) the distributions of errors of the GWO models are different in 4 of 10 cases; (2) the distributions of errors of the machine learning models differ from the distributions of errors of almost all the GWO models, except one case; (3) the distributions of errors of the machine learning models do not differ from each other.

Table 2.8 The exact *P*-statistics for the forecast models of the COVID-19 outbreak in Iran.

Method	LogitDWO	LinGWO	LogGWO	QGWO	PGWO	MLP	ANFIS
LogitDWO	1.000	0.964	0.893	0.929	0.893	0.857	0.929
LinGWO	0.964	1.000	0.893	0.964	0.964	0.893	0.893
LogGWO	0.893	0.893	1.000	0.893	0.893	0.893	0.893
QGWO	0.929	0.964	0.893	1.000	1.000	0.893	0.893
PGWO	0.893	0.964	0.893	1.000	1.000	0.857	0.857
MLP	0.857	0.893	0.893	0.893	0.857	1.000	0.964
ANFIS	0.929	0.893	0.893	0.893	0.857	0.964	1.000

Using the exact P-statistics for data on the COVID-19 outbreak in Iran, we may conclude that (1) the distributions of errors of the GWO models are different in 6 of 10 cases; (2) the distributions of errors of the machine learning models differ from the distributions of errors of all the GWO models; (3) the distributions of errors of the machine learning models do not differ from each other.

Table 2.9 The exact P-statistics for the forecast models of the COVID-19 outbreak in China.

Method	LogitDWO	LinGWO	LogGWO	QGWO	PGWO	MLP	ANFIS
LogitDWO	1.000	0.893	0.893	0.893	0.893	1.000	1.000
LinGWO	0.893	1.000	0.893	1.000	1.000	0.857	0.857
LogGWO	0.893	0.893	1.000	0.893	0.893	0.893	0.893
QGWO	0.893	1.000	0.893	1.000	1.000	0.857	0.857
PGWO	0.893	1.000	0.893	1.000	1.000	0.857	0.857
MLP	1.000	0.857	0.893	0.857	0.857	1.000	1.000
ANFIS	1.000	0.857	0.893	0.857	0.857	1.000	1.000

Using the exact P-statistics for data on the COVID-19 outbreak in China, we may conclude that (1) the distributions of errors of the GWO models are different in 7 of 10 cases; (2) the distributions of errors of the machine learning models differ from the distributions of errors of all the GWO models, except one case; (3) the distributions of errors of the machine learning models do not differ from each other.

Table 2.10 The exact P-statistics for the forecast models of the COVID-19 outbreak in the United States.

Method	LogitDWO	LinGWO	LogGWO	QGWO	PGWO	MLP	ANFIS
LogitDWO	1.0000	0.857	0.964	0.964	0.857	1.000	1.000
LinGWO	0.857	1.000	0.893	0.964	1.000	0.893	0.893
LogGWO	0.964	0.893	1.000	0.893	0.964	0.964	0.964
QGWO	0.964	0.964	0.893	1.000	0.964	0.893	0.893
PGWO	0.857	1.000	0.964	0.964	1.000	0.893	0.893
MLP	1.000	0.893	0.964	0.893	0.892	1.000	0.929
ANFIS	1.000	0.893	0.964	0.893	0.893	0.929	1.000

Using the exact P-statistics for the data on the COVID outbreak in the United States, we may conclude that (1) the distributions of errors of the GWO models are different in 4 of 10 cases; (2) the distributions of errors of the machine learning models differ from the distributions of errors of all the GWO models, except one case; (3) the distributions of errors of the machine learning models differ from each other.

confidence interval is wider than the Wilson interval. Thus, the exact test is less sensitive than the Klyushin–Petunin test, but has a guaranteed significance level. The fact that the distributions of model errors, mostly, are different allows for comparing these models by accuracy. As far as all conclusions in most cases are similar, the data obtained in the mentioned countries may be considered consistent.

Tables 2.11–2.15 show that the Kolmogorov–Smirnov test considered the models as statistically different in 83.4% of cases (when the confidence limit exceeded 0.95). Note that the Kolmogorov–Smirnov test considers the machine learning methods statistically equivalent. The GWO modifications have more variability. The fact that the distributions of model errors, mostly, are different allows for comparing these models by accuracy. As far as all conclusions in most cases are similar, the data obtained in the mentioned countries may be considered as consistent.

Table 2.11 The confidence level of the Kolmogorov–Smirnov test for the forecast models of the COVID-19 outbreak in Italy.

Method	LogitDWO	LinGWO	LogGWO	QGWO	PGWO	MLP	ANFIS
LogitDWO	1.000	0.480	1.000	0.992	0.992	0.872	0.872
LinGWO	0.480	1.000	1.000	0.992	0.992	0.480	0.480
LogGWO	1.000	0.992	1.000	1.000	1.000	1.000	1.000
QGWO	0.992	0.992	0.992	1.000	0.519	0.812	0.812
PGWO	0.002	0.972	1.000	0.519	1.000	0.972	0.812
MLP	0.972	0.972	1.000	0.812	0.812	1.000	0.812
ANFIS	0.972	0.972	1.000	0.812	0.812	0.812	1.000

Using the Kolmogorov–Smirnov test for the data on the COVID-19 outbreak in Italy, we may conclude that (1) the distributions of errors of the GWO models are different in 1 of 10 cases; (2) the distributions of errors of the machine learning models differ from the distributions of errors of all the GWO models, except one case; (3) the distributions of errors of the machine learning models differ from each other.

Table 2.12 The confidence level of the Kolmogorov–Smirnov test for the forecast models of the COVID-19 outbreak in Germany.

Method	LogitDWO	LinGWO	LogGWO	QGWO	PGWO	MLP	ANFIS
LogitDWO	1.000	0.999	0.999	0.941	0.999	0.480	0.812
LinGWO	0.999	1.000	1.000	0.992	0.906	0.951	0.999
LogGWO	0.999	1.000	1.000	1.000	0.999	0.951	0.999
QGWO	0.941	0.992	1.000	1.000	0.992	0.951	0.960
PGWO	0.999	0.906	0.999	0.992	1.000	0.951	0.999
MLP	0.480	0.951	0.951	0.951	0.951	1.000	0.480
ANFIS	0.812	0.999	0.999	0.960	0.999	0.480	1.000

Using the Kolmogorov–Smirnov test for data on COVID-19 outbreak in Germany, we may conclude that (1) the distributions of errors of the GWO models are different in 1 of 10 cases; (2) the distributions of errors of the machine learning models differ from the distributions of errors of all the GWO models, except one case; (3) the distributions of errors of the machine learning models differ from each other.

Table 2.13 The confidence level of the Kolmogorov–Smirnov test for the forecast models of the COVID-19 outbreak in Iran.

Method	LogitDWO	LinGWO	LogGWO	QGWO	PGWO	MLP	ANFIS
LogitDWO	1.000	0.992	1.000	0.999	0.999	0.951	0.951
LinGWO	0.992	1.000	1.000	0.951	0.992	0.992	0.992
LogGWO	1.000	0.951	1.000	1.000	1.000	1.000	1.000
QGWO	0.999	0.992	1.000	1.000	0.812	0.999	0.999
PGWO	0.999	0.992	1.000	0.812	1.000	0.999	0.999
MLP	0.951	0.992	1.000	0.999	0.999	1.000	0.951
ANFIS	0.951	0.992	1.000	0.999	0.999	0.951	1.000

Using the Kolmogorov–Smirnov test for the data on the COVID-19 outbreak in Iran, we may conclude that (1) the distributions of errors of the GWO models are different in 1 of 10 cases; (2) the distributions of errors of the machine learning models differ from the distributions of errors of all the GWO models, except one case; (3) the distributions of errors of the machine learning models differ from each other.

Table 2.14 The confidence level of the Kolmogorov–Smirnov test for the forecast models of the COVID-19 outbreak in China.

Method	LogitDWO	LinGWO	LogGWO	QGWO	PGWO	MLP	ANFIS
LogitDWO	1.000	0999	1.000	0.999	0.999	0.480	0.812
LinGWO	0.999	1.000	1.000	0.812	0.812	0.999	0.999
LogGWO	1.000	1.000	1.000	1.000	1.000	1.000	1.000
QGWO	0.999	0.999	1.000	1.000	1.000	0.999	0.999
PGWO	0.999	0.999	0.812	1.000	1.000	0.999	0.999
MLP	0.480	0.480	0.999	0.999	0.999	1.000	0.812
ANFIS	0.812	0.812	0.999	0.999	0.999	0.812	1.000

Using the Kolmogorov–Smirnov test for the data on the COVID-19 outbreak in China, we may conclude that (1) the distributions of errors of the GWO models are different in 2 of 10 cases; (2) the distributions of errors of the machine learning models differ in 3 of 5 GWO models; (3) the distributions of errors of machine learning models differ from each other.

Table 2.15 The confidence level of the Kolmogorov–Smirnov test for the forecast models of the COVID outbreak in the United States.

Method	LogitDWO	LinGWO	LogGWO	QGWO	PGWO	MLP	ANFIS
LogitDWO	1.000	0.999	0.999	0.951	0.999	0.951	0.872
LinGWO	0.999	1.000	1.000	0.992	0.906	0.999	0.999
LogGWO	0.999	0.999	1.000	1.000	0.999	0.999	0.999
QGWO	0.951	0.951	1.000	1.000	0.991	0.951	0.951
PGWO	0.999	0.999	0.999	0.991	1.000	0.999	0.999
MLP	0.951	0.951	0.999	0.951	0.999	1.000	0.951
ANFIS	0.862	0.872	0.999	0.951	0.999	0.951	1.000

Using the Kolmogorov–Smirnov test for the data on COVID outbreak in the United States, we may conclude that (1) the distributions of errors of GWO models are different in 2 of 10 cases; (2) the distributions of errors of the machine learning models differ from the distributions of errors of the GWO models in three of five cases; (3) the distributions of errors of the machine learning models differ from each other.

Tables 2.16–2.20 demonstrate that the sign test considers the models as statistically equivalent (confidence level is less than 0.95) in 9 cases from 21 for Italy, 9 cases for Germany, 9 cases for Iran, 10 cases for China, and 10 cases for the United States. Thus, it distinguished the models in 44.7% of cases. Also, we may conclude that the data from all considered countries are consistent. From the point of view of the sign test, the machine learning methods produced statistically equivalent results in all cases of comparisons. The versions of the GWO methods also, as a rule, produced statistically different results. In general, most comparisons using considered nonparametric tests demonstrate the homogeneity of the results. However, the tests have different sensitivity. The most sensitive is the

Table 2.16 Confidence level of the sign test for the forecast models of the COVID-19 outbreak in Italy.

Method	LogitDWO	LinGWO	LogGWO	QGWO	PGWO	MLP	ANFIS
LogitDWO	1.000	0.453	0.992	0.937	0.962	0.156	0.600
LinGWO	0.453	1.000	0.992	0.884	0.884	0.453	0.453
LogGWO	0.992	0.992	1.000	0.992	0.992	0.992	0.992
QGWO	0.937	0.884	0.992	1.000	0.961	0.967	0.967
PGWO	0.962	0.884	0.992	0.961	1.000	0.961	0.961
MLP	0.156	0.453	0.992	0.961	0.961	1.000	0.453
ANFIS	0.600	0.453	0.992	0.961	0.961	0.453	1.000

Using the sign test for the data on COVID-19 outbreak in Italy, we may conclude that (1) the distributions of errors of the GWO models are different in 5 of 10 cases; (2) the distributions of errors of the machine learning models differ from the distributions of errors of the GWO models in three of five cases; (3) the distributions of errors of the machine learning models differ from each other.

Table 2.17 Confidence level of the sign test for the forecast models of the COVID-19 outbreak in Germany.

Method	LogitDWO	LinGWO	LogGWO	QGWO	PGWO	MLP	ANFIS
LogitDWO	1.000	0.984	0.992	0.852	0.992	0.258	0.258
LinGWO	0.984	1.000	0.992	0.984	0.844	0.844	0.844
LogGWO	0.992	0.992	1.000	0.992	0.992	0.992	0.992
QGWO	0.852	0.984	0.992	1.000	0.992	0.852	0.853
PGWO	0.992	0.844	0.992	0.992	1.000	0.992	0.992
MLP	0.258	0.884	0.992	0.852	0.992	1.000	0.453
ANFIS	0.258	0.884	0.992	0.853	0.992	0.453	1.000

Using the sign test for data on the COVID-19 outbreak in Germany, we may conclude that (1) the distributions of errors of the GWO models are different in 4 of 10 cases; (2) the distributions of errors of machine learning models differ from the distributions of errors of the GWO models in three of five cases; (3) the distributions of errors of the machine learning models differ from each other.

original Klyushin–Petunin test (84.7%), followed by the Kolmogorov–Smirnov (83.4%), the exact test (80.9%), and the sign test (44.7%).

To illustrate the effectiveness of the method described above for other data and models, consider random forest model (RFM), K-nearest neighbor (KNN) model, and gradient boosting model (GBM) forecasting the number of cases of the COVID-19 in Germany, Japan, South Korea, and Ukraine in February–April 2022 [60]. As benchmarks, we used the Diebold–Mariano and the Wilcoxon signed-rank tests. The null hypothesis that the distributions are identical is rejected if the P-statistics is greater than 0.95 and the P-values of the Diebold–Mariano and Wilcoxon signed-rank tests are greater than 0.05.

In Table 2.21, the P-statistics and the P-value of the Wilcoxon signed-rank tests are constant because the samples of the absolute errors of all the models are

Table 2.18 Confidence level of the sign test for the forecast models of the COVID-19 outbreak in Iran.

Method	LogitDWO	LinGWO	LogGWO	QGWO	PGWO	MLP	ANFIS
LogitDWO	1.000	0.992	0.992	0.992	0.884	0.336	0.852
LinGWO	0.992	1.000	0.992	0.884	0.884	0.992	0.992
LogGWO	0.992	0.992	1.000	0.992	0.992	0.992	0.992
QGWO	0.992	0.884	0.992	1.000	0.884	0.992	0.992
PGWO	0.884	0.884	0.992	0.884	1.000	0.884	0.884
MLP	0.336	0.992	0.992	0.992	0.884	1.000	0.258
ANFIS	0.852	0.992	0.992	0.992	0.884	0.258	1.000

Using the sign test for the data on the COVID-19 outbreak in Iran, we may conclude that (1) the distributions of errors of the GWO models are different in 4 of 10 cases; (2) the distributions of errors of the machine learning models differ from the distributions of errors of the GWO models in two of five cases; (3) the distributions of errors of the machine learning models differ from each other.

Table 2.19 Confidence level of the sign test for the forecast models of the COVID-19 outbreak in China.

Method	LogitDWO	LinGWO	LogGWO	QGWO	PGWO	MLP	ANFIS
LogitDWO	1.000	0.884	0.992	0.884	0.884	0.957	0.932
LinGWO	0.884	1.000	0.992	0.884	0.884	0.884	0.884
LogGWO	0.992	0.992	1.000	0.992	0.992	0.992	0.992
QGWO	0.884	0.884	0.992	1.000	0.852	0.984	0.984
PGWO	0.884	0.884	0.992	0.852	1.000	0.984	0.984
MLP	0.957	0.884	0.992	0.984	0.984	1.000	0.932
ANFIS	0.953	0.884	0.992	0.984	0.984	0.932	1.000

Using the sign test for the data on the COVID-19 outbreak in China, we may conclude that (1) the distributions of errors of the GWO models are different in 6 of 10 cases; (2) the distributions of errors of the machine learning models differ from the distributions of errors of the GWO models in four of five cases; (3) the distributions of errors of the machine learning models differ from each other.

Table 2.20 Confidence level of the sign test for the forecast models of the COVID-19 outbreak in the United States.

Method	LogitDWO	LinGWO	LogGWO	QGWO	PGWO	MLP	ANFIS
LogitDWO	1.000	0.884	0.992	0.852	0.884	0.805	0.805
LinGWO	0.884	1.000	0.992	0.984	0.992	0.984	0.984
LogGWO	0.992	0.992	1.000	0.992	0.992	0.992	0.992
QGWO	0.852	0.984	0.992	1.000	0.992	0.955	0.852
PGWO	0.84	0.992	0.992	0.992	1.000	0.884	0.884
MLP	0.805	0.984	0.992	0.955	0.884	1.000	0.805
ANFIS	0.805	0.984	0.992	0.852	0.884	0.805	1.000

Using the sign test for the data on the COVID-19 outbreak in the United States, we may conclude that (1) the distributions of errors of the GWO models are different in 4 of 10 cases; (2) the distributions of errors of the machine learning models differ from the distributions of errors of the GWO models in four of five cases; (3) the distributions of errors of the machine learning models differ from each other.

Table 2.21 P-statistics and P-values of the Klyushin–Petunin, Diebold–Mariano, and Wilcoxon signed-rank tests for prediction of the COVID-19 cases in Germany, Japan, South Korea, and Ukraine using random forest model (RFM), K-nearest neighbor (KNN) model, and gradient boosting model (GBM).

Tests/methods	P-statistics of the Klyushin–Petunin test			P-value of the Diebold–Mariano test			P-value of the Wilcoxon signed-rank test		
	RFM	KNN	GBM	RFM	KNN	GBM	RFM	KNN	GBM
Germany									
RFM	1.0000	0.9301	0.9301	1.0000	0.0239	0.0161	1.0000	0.0025	0.0025
KNN	—	1.0000	0.9301	—	1.0000	0.1946	—	1.0000	0.0025
GBM	—	—	1.0000	—	—	1.0000	—	—	1.0000
Japan									
RFM	1.0000	0.9301	0.9301	1.000	1.943×10^{-5}	1.951×10^{-5}	1.0000	0.0020	0.0025
KNN	—	1.0000	0.9301	—	1.0000	1.241×10^{-5}	—	1.0000	0.0025
GBM	—	—	1.0000	—	—	1.0000	—	—	1.0000
South Korea									
RFM	1.0000	0.9301	0.9301	0.0000	2.371×10^{-4}	2.928×10^{-5}	1.0000	0.0020	0.0025
KNN	—	1.0000	0.9301	—	0.0000	1.445×10^{-2}	—	1.0000	0.0025
GBM	—	—	1.000	—	—	0.0000	—	—	1.0000
Ukraine									
RFM	1.0000	0.9301	0.9301	0.0000	2.371×10^{-4}	2.928×10^{-5}	1.0000	0.0020	0.0025
KNN	—	1.0000	0.9301	—	0.0000	1.445×10^{-2}	—	1.0000	0.0025
GBM	—	—	1.0000	—	—	0.0000	—	—	1.0000

not overlapping. Moreover, since the *P*-statistics is less than 0.95, and almost all the *P*-values of the Diebold—Mariano and the Wilcoxon signed-rank tests are less than 0.05, almost all the forecasting models for Germany, Japan, South Korea, and Ukraine are considered as different with the significance level $\alpha = 0.05$. The unique case when the Diebold—Mariano test cannot detect the difference is the case of the KNN model and the gradient boosting model for data from Germany. Therefore, the performances of the Klyushin—Petunin, Diebold—Mariano, and Wilcoxon signed-rank tests in this investigation are the same.

2.5 Conclusion and scope for the future work

Comparing the accuracy of the models, it is necessary to assess the statistical homogeneity of their errors. This problem is solved by the use of two-sample methods for checking the homogeneity of samples (the Klyushin—Petunin test, the Kolmogorov-Smirnov test, the sign test, etc.). The chapter describes the results of applying four nonparametric tests to compare the results of time-series forecasts (the Klyushin—Petunin test, the proposed exact nonparametric test, the Kolmogorov—Smirnov test, and the sign test). The results of comparing machine learning prediction models for different countries show statistical heterogeneity of errors (for three tests of four this sensitivity exceeds 80%). Thus, when ranking these models, we can really rely on the absolute values of their mean values. Also, the results demonstrate the high performance of the proposed exact nonparametric test. Its accuracy is comparable with the Klyushin—Petunin test, but, in addition, it has a guaranteed significance level; i.e., the significance level never exceeds 0.05.

Different versions of the GWO methods generate errors with different distributions depending on the underlying analytical dependencies used. In most cases, the MLP and ANFIS are classified as equivalent, so the choice of the most appropriate model should be made taking into account additional factors, such as computational complexity. The MLP has proven to be the most accurate, and therefore, it can be recommended as the main tool for predicting the COVID-19 outbreak. Using the Klyushin—Petunin, Diebold—Mariano, and Wilcoxon tests to test the hypothesis about the identity of error distributions of three COVID-19 forecasting models for Germany, Japan, South Korea, and Ukraine (February—April 2022), we have demonstrated that the performances of all the considered tests in this investigation are the same, and the gradient boosting model has the best precision. The prospects of the described work are related to the application of new models for forecasting the COVID-19 pandemic.

References

[1] S. Eker, Validity and usefulness of COVID-19 models, Humanit. Soc. Sci. Commun. 7 (54) (2020). Available from: https://doi.org/10.1057/s41599-020-00553-4.

[2] H.W. Hethcote, The mathematics of infectious diseases, SIAM Rev. 42 (4) (2000) 599−653. Available from: https://doi.org/10.1137/S0036144500371907.

[3] W.O. Kermack, A.G. McKendrick, A contribution to the mathematical theory of epidemics, Proc. R. Soc. Lond. A 115 (1927) 700−721. Available from: https://doi.org/10.1098/rspa.1927.0118.

[4] J. Aron, I. Schwartz, Seasonality and period-doubling bifurcations in an epidemic model, J. Theor. Biol. 110 (1984) 665−679. Available from: https://doi.org/10.1016/S0022-5193(84)80150-2.

[5] R. Li, S. Pei, B. Chen, Y. Song, T. Zhang, W. Yang, et al., Substantial undocumented infection facilitates the rapid dissemination of novel coronavirus (SARS-CoV-2), Science 368 (6490) (2020) 489−493. Available from: https://doi.org/10.1111/jebm.12376.

[6] N. Anand, A. Sabarinath, S. Geetha, S. Somanath, Predicting the spread of COVID-19 using SIR model augmented to incorporate quarantine and testing, Trans. Indian Natl Acad. Eng. 5 (2020) 141−148. Available from: https://doi.org/10.1007/s41403-020-00151-5.

[7] O. Ifguis, M.E. Ghozlani, F. Ammou, A. Moutcine, Z. Abdellah, Simulation of the final size of the evolution curve of coronavirus epidemic in Morocco using the SIR model, J. Environ. Public Health 2020 (2020). Available from: https://doi.org/10.1155/2020/9769267. Article ID 9769267.

[8] I. Nesteruk, Statistics-based predictions of coronavirus epidemic spreading in mainland China, Innov. Biosyst. Bioeng. 4 (1) (2020) 13−18. Available from: https://doi.org/10.20535/ibb.2020.4.1.195074.

[9] M. Babu, S. Marimuthu, M. Joy, A. Nadaraj, E. Asirvatham, L. Jeyaseelan, Forecasting COVID-19 epidemic in India and high incidence states using SIR and logistic growth models, Clin. Epidemiol. Glob. Health 9 (2020) 26−33. Available from: https://doi.org/10.1016/j.cegh.2020.06.006.

[10] J. He, G. Chena, Y. Jiang, R. Jin, R. Shortridge, S. Agusti, et al., Comparative infection modeling and control of COVID-19 transmission patterns in China, South Korea, Italy and Iran, Sci. Total Environ. 74 (141447) (2020). Available from: https://doi.org/10.1016/j.scitotenv.2020.141447.

[11] A. Guirao, The Covid-19 outbreak in Spain. A simple dynamics model, some lessons, and a theoretical framework for control response, Infect. Dis. Model. 5 (2020) 652−669. Available from: https://doi.org/10.1016/j.idm.2020.08.010.

[12] N. Wang, Y. Fu, H. Zhang, H. Shi, An evaluation of mathematical models for the outbreak of COVID-19, Precis. Clin. Med. 3 (2) (2020) 85−93. Available from: https://doi.org/10.1093/pcmedi/pbaa016.

[13] H. Swapnarekha, H.S. Behera, J. Nayak, B. Naik, Role of intelligent computing in COVID-19 prognosis: a state-of-the-art review, Chaos Solitons Fract. (2020) 109947. Available from: https://doi.org/10.1016/j.chaos.2020.109947.

[14] R. Sujath, J.M. Chatterjee, A.E. Hassanien, A machine learning forecasting model for COVID-19 pandemic in India, Stoch. Environ. Res. Risk Assess. (2020). Available from: https://doi.org/10.1007/s00477-020-01827-8.

[15] S.F. Ardabili, A. Mosavi, P. Ghamisi, F. Ferdinand, A.R. Varkonyi-Koczy, U. Reuter, et al., COVID-19 Outbreak Prediction with Machine Learning, Preprints 2020040311, (2020). doi:10.20944/preprints202004.0311.v1.

[16] S. Tuli, S. Tuli, R. Tuli, S.S. Gill, Predicting the growth and trend of COVID-19 pandemic using machine learning and cloud computing, Internet Things 11 (100222) (2020). Available from: https://doi.org/10.1016/j.iot.2020.100222.

[17] C. Distante, I. Pereira, L. Gonçalves, P. Piscitelli, A. Miani, Forecasting Covid-19 outbreak progression in Italian regions: a model based on neural network training from Chinese data, MedRxiv (2020). Available from: https://doi.org/10.1101/2020.04.09.20059055.

[18] L.R. Kolozsvári, T. Bérczes, A. Hajdu, R. Gesztelyi, A. Tiba, I. Varga, et al., Predicting the epidemic curve of the coronavirus (SARS-CoV-2) disease (COVID-19) using artificial intelligence, MedRxiv (2020). Available from: https://doi.org/10.1101/2020.04.17.20069666.

[19] M.R. Ibrahim, J. Haworth, A. Lipani, A. Aslam, T. Cheng, N. Christie, Variational-LSTM Autoencoder to forecast the spread of coronavirus across the globe, MedRxiv (2020). Available from: https://doi.org/10.1101/2020.04.20.20070938.

[20] Z. Hu, Q. Ge, S. Li, E. Boerwincle, L. Jin, M. Xiong, Forecasting and evaluating multiple interventions of Covid-19 worldwide, Front. Artif. Intell. 2020 (2020) 00041. Available from: https://doi.org/10.3389/frai.2020.00041.

[21] Q. Guo, Z. He, Prediction of the confirmed cases and deaths of global COVID-19 using artificial intelligence, Environ. Sci. Pollut. Res. 28 (2021) 11672−11682. Available from: https://doi.org/10.1007/s11356-020-11930-6.

[22] S. Ballı, Data analysis of Covid-19 pandemic and short-term cumulative case forecasting using machine learning time series methods, Chaos Solitons Fract. 142 (2021) 110512. Available from: https://doi.org/10.1016/j.chaos.2020.110512. Jan.

[23] R. Kafieh, R. Arian, N. Saeedizadeh, Z. Amini, N.D. Serej, S. Minaee, et al., COVID-19 in Iran: forecasting pandemic using deep learning, Comput. Math. Methods Med. (2021). Available from: https://doi.org/10.1155/2021/6927985. Article ID 6927985.

[24] F. Rustam, et al., COVID-19 future forecasting using supervised machine learning models, IEEE Access. 8 (2020) 101489−101499. Available from: https://doi.org/10.1109/ACCESS.2020.2997311.

[25] F.X. Diebold, R.S. Mariano, Comparing predictive accuracy, J. Bus. Econ. Stat. 13 (1995) 253−263. Available from: https://doi.org/10.1080/07350015.1995.10524599.

[26] B.E. Flores, The utilization of the Wilcoxon test to compare forecasting methods: a note, Int. J. Forecast. 5 (1989) 529−535. Available from: https://doi.org/10.1016/0169-2070(89)90008-3.

[27] T. DelSole, M.K. Tippett, Comparing forecast skill, Mon. Weather Rev. 142 (2014) 4658−4678. Available from: https://doi.org/10.1175/MWR-D-14-00045.1.

[28] S. Mirjalili, S.M. Mirjalili, A. Lewis, Grey wolf optimizer, Adv. Eng. Softw. 69 (2014) 46−61. Available from: https://doi.org/10.1016/j.advengsoft.2013.12.007.

[29] J.S. Jang, ANFIS: adaptive-network-based fuzzy inference system, IEEE Trans. Syst. Man Cybern. 23 (1993) 665−685. Available from: https://doi.org/10.1109/21.256541.

[30] N.V. Smirnov, Estimate of difference between empirical distribution curves in two independent samples, Bull. Mosk. Gos. Univ. 2 (2) (1939) 3−14.

[31] N.V. Smirnov, On the deviations of an empirical distribution curve, Mat. Sb. 6 (1) (1939) 3−26.

[32] W.G. Dixon, A criterion for testing the hypothesis that two samples are from the same population, Ann. Math. Stat. 11 (1940) 199−204. Available from: https://doi.org/10.1214/AOMS/1177731914.

[33] A. Wald, J. Wolfowitz, On a test whether two samples ate from the same population, Ann. Math. Stat. 11 (1940) 147−162. Available from: https://doi.org/10.1214/AOMS/1177731909.

[34] H.C. Mathisen, A method of testing the hypothesis that two samples are from the same population, Ann. Math. Stat. 14 (1943) 188−194. Available from: https://doi.org/10.1214/aoms/1177731460.

[35] F. Wilcoxon, Individual comparisons by ranking methods, Biometrika 1 (1945) 80−83. Available from: https://doi.org/10.2307/3001968.

[36] H.B. Mann, D.R. Whitney, On a test of whether one of the random variables is stochastically larger than other, Ann. Math. Stat. 18 (1947) 50−60. Available from: https://doi.org/10.1214/aoms/1177730491.

[37] S.S. Wilks, A combinatorial test for the problem of two samples from continuous distributions, Proceedings of the Fourth Berkeley Symposium on Mathematical Statistics and Probability, Vol. 1, 1961, pp. 707−717.

[38] E.J.G. Pitman, Significance tests which may be applied to samples from any populations, J. R. Stat. Soc. Ser. A. 4 (1937) 119−130. Available from: https://doi.org/10.2307/2983647.

[39] E.L. Lehmann, Consistency and unbiasedness of certain nonparametric tests, Ann. Math. Stat. 22 (1947) 165−179. Available from: https://doi.org/10.1214/aoms/1177729639.

[40] M. Rosenblatt, Limit theorems associated with variants of the von Mises statistic, Ann. Math. Stat. 23 (1952) 617−623. Available from: https://doi.org/10.1214/aoms/1177729341.

[41] M. Dwass, Modified randomization tests for nonparametric hypotheses, Ann. Math. Stat. 28 (1957) 181−187. Available from: https://doi.org/10.1214/aoms/1177707045.

[42] M. Fisz, On a result be M. Rosenblatt concerning the Mises−Smirnov test, Ann. Math. Stat. 31 (1960) 427−429. Available from: https://doi.org/10.1214/aoms/1177705905.

[43] G.A. Barnard, Comment on "The spectral analysis of point processes" by M.S. Bartlett, J. R. Stat. Soc. Ser. B 25 (1963) 294.

[44] Z.W. Birnbaum, Computers and unconventional test-statistics, in: F. Prochan, R.J. Serfling (Eds.), Reliability and Biometry, SIAM, Philadelphia, PA, 1974, pp. 441−458.

[45] K.H. Jockel, Finite sample properties and asymptotic efficiency of Monte Carlo tests, Ann. Stat. 14 (1986) 336−347. Available from: https://doi.org/10.1214/aos/1176349860.

[46] D.L. Allen, Hypothesis testing using L1-distance bootstrap, Am. Stat. 51 (1997) 145−150. Available from: https://doi.org/10.1080/00031305.1997.10473949.

[47] B. Efron, R.J. Tibshirani, An introduction to the bootstrap, Vol. 57 of Monographs on Statistics and Applied Probability, Chapman-Hall, New York, 1993.

[48] J.-M. Dufour, A. Farhat, Exact nonparametric two-sample homogeneity tests for possibly discrete distributions, Center for Interuniversity Research in Quantitative Economics (CIREQ). Preprint 2001-23, California Press, 2001, pp. 707−717.

[49] H. Cramér, On the composition of elementary errors, Scand. Actuar. J. (1)(1928) 13−74. Available from: https://doi.org/10.1080/03461238.1928.10416862.

[50] R.E. von Mises, Wahrscheinlichkeit, Statistik und Wahrheit, Julius Springer, 1928.

[51] T.W. Anderson, On the distribution of the two-sample Cramer-von Mises criterion, Ann. Math. Stat. (33)(1962) 1148−1159. Available from: https://doi.org/10.1214/aoms/1177704477.

[52] D.A. Darling, The Kolmogorov-Smirnov, Cramer-von Mises tests, Ann. Math. Stat. (28)(1957) 223−238. Available from: https://doi.org/10.1214/aoms/1177728589.

[53] B.M. Hill, Posterior distribution of percentiles: Bayes' theorem for sampling from a population, J. Am. Stat. Assoc. 63 (1968) 677−691. Available from: https://doi.org/10.1080/01621459.1968.11009286.

[54] S.A. Matveichuk SA, Y.I. Petunin, Generalization of Bernoulli schemes that arise in order statistics, I. Ukrainian Math. J. 42 (4) (1990) 459−466. Available from: https://doi.org/10.1007/BF01071335.

[55] S.A. Matveichuk, Y.I. Petunin, Generalization of Bernoulli schemes that arise in order statistics. II, Ukrainian Math. J. 43 (6) (1991) 728−734. Available from: https://doi.org/10.1007/BF01058940.

[56] N. Johnson, S. Kotz, Some generalizations of Bernoulli and Polya-Eggenberger contagion models, Stat. Pap. 32 (1991) 1−17. Available from: https://doi.org/10.1007/BF02925473.

[57] A.M. Pires, C. Amado, Interval estimators for a binomial proportion: comparison of twenty methods, REVSTAT−Stat. J. (6)(2008) 165−197.

[58] R.I. Andrushkiw, D.A. Klyushin, Yu.I. Petunin, M. Yu Savkina, The exact confidence limits for unknown probability Bernoulli models, Proceedings of the International Conference on Information Technology Interfaces, ITI, 2005, pp. 164−168. Available from: http://doi.org/10.1109/ITI.2005.1491116.

[59] D.A. Klyushin, Y.I. Petunin, A nonparametric test for the equivalence of populations based on a measure of proximity of samples, Ukrainian Math. J. 55 (2) (2003) 181−198. Available from: https://doi.org/10.1023/A:1025495727612.

[60] D. Chumachenko, I. Meniailov, L. Bazilevych, T. Chumachenko, S. Yakovlev, Investigation of statistical machine learning models for COVID-19 epidemic process simulation: random forest, K-nearest neighbors, gradient boosting, Computation 10 (86) (2022). Available from: https://doi.org/10.3390/computation10060086.

Reconsideration of drug repurposing through artificial intelligence program for the treatment of the novel coronavirus

Lakshmi Narasimha Gunturu[1], Girirajasekhar Dornadula[2] and Raghavendra Naveen Nimbagal[3]

[1]*Scientimed Solutions Private Limited, Mumbai, Maharashtra, India*
[2]*Department of Pharmacy Practice, Annamacharya College of Pharmacy, Rajampeta, Andhra Pradesh, India*
[3]*Department of Pharmaceutics, Sri Adichunchanagiri College of Pharmacy, Adichunchanagiri University, B.G. Nagar, Mandya, Karnataka, India*

3.1 Introduction

The eruption of the coronavirus termed severe acute respiratory syndrome coronavirus (SARS-CoV-2) in Wuhan, China, made the public panic. Since its outburst and rapid transmission among the public, World Health Organization (WHO) has declared the condition a pandemic. The mode of transmission of coronavirus is mainly by the respiratory particles as well as from the airborne droplets [1,2]. Humans exposed to this virus have the chance of developing the symptoms after 2 weeks or approximately 14 days and sometimes even can remain asymptomatic. The common symptoms so far reported in patients affected with novel coronavirus are high body temperature, fatigue, dry cough, and acute respiratory distress syndrome. Apart from those clinical symptoms, death cases are also reported in some patients in the case of elderly patients. For these reasons, COVID-19 poses a great danger to the public world. Till January 16, 2021, 94,414,806 active cases have been reported among the different countries, with the United States, India, Brazil, Russia, and the United Kingdom in the top five on the list of being highly infected by the virus [3,4].

International Committee on Taxonomy (ICT), which is accountable for providing nomenclature to the viruses based on their phylogeny, had named this novel coronavirus 2019 (2019-nCoV) SARS-CoV-2 [5]. Since its outbreak, no specific drug therapy has existed to treat this virus in therapeutic areas. So far, drugs like antivirals, antibiotics, and other immune modulators used in the treatment only provide symptomatic relief [6]. To minimize the viral spread among the public, government recommended

Artificial Intelligence in Healthcare and COVID-19. DOI: https://doi.org/10.1016/B978-0-323-90531-2.00009-6

various guidelines like quarantine rules, social distance maintenance between the people, avoiding public gatherings, and tracking of the public with various apps and tracers [7]. Many therapeutic regimens and social and supportive measures utilized to avert the proliferation of novel coronavirus in use today are approaches based upon the evidence of previous pandemics like Middle East respiratory syndrome (MERS) and severe acute respiratory syndrome (SARS). Therefore, it is necessary to identify the lead molecules and efficient drug targets to tackle the novel coronavirus.

Due to prevailing conditions and a threat from the virus, current public experiences many challenges to protect themselves from the new coronavirus. Current technology approaches like computational intelligence play an important role in identifying new targets for drugs and vaccines that are efficient against the novel coronavirus. The possible implementation of computational intelligence comprises the identification of molecular structures of receptors linked with the disease advancement to recognize the importance of existing drugs in the therapeutic areas that are capable of targeting the disease-related genes or their protein structures (drug repurposing/drug reconsideration), machine learning mechanisms of active compounds identification, and offering new molecular entities or lead compounds that are effective against the inhibition of disease progression [8]. Therefore, studies are in the way to identify the new therapeutic targets against SARS-CoV-2. So, from a computational intelligence perspective, methods like molecular docking, network models, phenotype screening, and experimental approaches play a crucial role as drug repurposing strategies to find the drug molecule against the SARS-CoV-2 infection [9,10].

Amid various computational approaches, drug repurposing and utilizing existing or shelved drugs is a primary and efficient strategy for producing therapeutic compounds against the novel coronavirus. Drugs developed by this process had a better market success rate in contrast to the conventional approach (de novo drug synthesis). It is also an efficient approach to cutting the time and expenses associated with developing new therapeutic molecules. The utilization of drug repurposing approaches helps to answer various clinical, safety, therapeutic, and manufacturing-related problems, stability, and storage issues as they are already widely used and readily available in the market. Fig. 3.1 explains the differentiation between the conventional synthesis approach and the drug repurposing approach concerning their various phases and timelines.

The following sections deal with basic concepts on viral morphology and its life cycle, currently available viral drug candidates, different drug repurposing approaches, artificial intelligence (AI) algorithms for drug repurposing, computational intelligence-based therapeutic candidates repurposed against COVID-19, future challenges, and conclusions.

3.2 Viral morphology

Structural proteins associated with COVID-19 are spike membrane, nucleic acid membrane, a sheath of protein, and its envelope membrane [1,11].

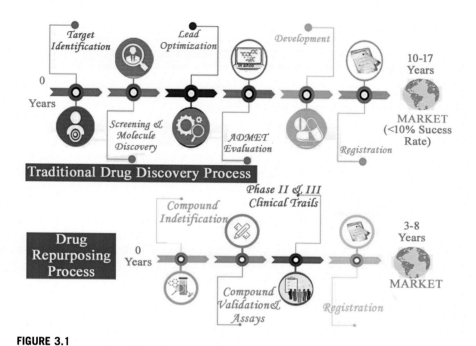

FIGURE 3.1

Differences between the traditional drug discovery method and drug repurposing.

3.2.1 Structured proteins

3.2.1.1 Spike protein/spike membrane

The outermost surface of coronavirus consists of spike proteins (S). They facilitate attachment of the virus, identification of host receptors, and penetration into the cell. This protein sheath is arranged in the form of a homotrimer. The protein sheath consists of the extracellular domain, the intracellular domain, and the transmembranous sheath. The outer region of spike protein consists of the receptor binding site (RBS), S-terminal, and N-terminals. Once SARS-CoV-2 enters into host cell, it binds to the surface of ACE 2 receptors by its RBS located at the S1 protein. Hence, in the drug repurposing strategies, new therapeutic molecules that can hamper the virus's binding to the host's cell surface are realized as potential target sites [12−14].

3.2.1.2 Membranous proteins

The structure of the virus depends upon the membranous proteins (M) and their interaction with other structural proteins. The surface of membranous proteins consists of a short N-terminal and a long S-terminal. They support the viral binding to cell receptors. In the cases of viral infection, there might be a chance of high M proteins in the collected sample, which aid in the virus's maturation. S terminal interactivity is needed especially for attachment of spike membranes with Golgi body, whereas N-terminal of membranous proteins interaction is needed to form viral core material [15].

3.2.1.3 Nucleic acid—protein/nucleocapsid

The vital components of nucleic acid—protein are the N-terminal, a central linker, and a C end. C tail is further referred to as an internally disordered site. The nucleic acid—protein of the virus contributes an essential role in the RNA binding mechanism. Central linker areas consist of amino acids, arginine and serine, that act as phosphorylation sites. C terminal plays a crucial part in the oligomerization process. The primary function of a nucleocapsid is to pack all the RNA molecules into ribonucleoprotein complexes and, in turn, convert them into capsid. This is termed self-assembly. It also plays a pivotal role in the viral life cycle by modulating host cellular activities. Therefore, drug repurposing methods target these nucleocapsid protein structures that might affect the virus life cycle [16—18].

3.2.1.4 Enveloped protein

This constitutes the smallest genomic size of nearly 8.4—12 kDa. It possesses mainly two components: the hydrophobic domain and the cytoplasmic tail end. Enveloped proteins help in the formation of viral structures [19].

3.2.2 Nonstructured proteins

These encode for the genetic material of the virus and help in the production of viral enzymes. These include protease, RNA-dependent polymerase, and helicases.

3.2.2.1 Proteases

The viral genetic material is responsible for the synthesis of many enzymes. Among them, proteases are one of the essential enzymes that contribute to a crucial role in the virus life cycle. In the nonstructural protein (NSP) part, nearly 16 NSPs are responsible for performing various functions in the SARS-CoV-2. The list of all those proteins is provided in Table 3.1. All these NSPs exist as polyproteins. The enzyme proteinase is present in two forms: C-like proteinase and papain (P)-like proteinase. This enzyme was responsible for breaking bonds between the large polyproteins and helping synthesize NSP. In the formation of 16 NSPs, first three are formed by the cleavage of P-like proteinase, and the rest 11 are formed by the C-like proteinases. C-like proteinases help carry out cellular processes like transcription, translation, and replication, whereas P-like proteinases are needed to break down replicase substrate [20,21].

Table 3.1 Different nonstructural proteins and their activities.

No.	Nonstructural proteins associated with COVID-19	Functions of nonstructural proteins
1	NSP1	Helps in mRNA degradation
2	NSP2	Attaches to prohibiting proteins
3	NSP3	Initiates cytokine response and inhibits host immune response
4	NSP4	Needed for structure formation
5	NSP5	It breaks viral proteins.
6	NSP6	Terminates autophagosome activity
7	NSP7	Assists in the process of RNA polymerase
8	NSP8	Acts as cofactor with NSP7 and NSP12
9	NSP9	RNA binding and dimerization process
10	NSP10	Cofactor for NSP16 and NSP14
11	NSP12	Promotes transcription of negative stranded RNA.
12	NSP13	Break down of all types of nucleoside triphosphate hydrolases
13	NSP14	Replication of SARS-CoV-2
14	NSP15	Viral endonuclease protein
15	NSP16	Releases viral RNA and alters host immunity

3.2.2.2 RNA-dependent polymerase

This is also referred to as NSP12 and takes part in the significant replication process of SARS-CoV-2. The ORF-1 gene codes it at three prime ends. Among the therapeutic candidates acting against the coronavirus, those acting on RNA-dependent polymerase can have an essential role in preventing viral spread along with a reduced toxicity profile in the host [22].

3.2.2.3 Helicase

It serves as a multifunctional protein. The N-terminal site of helicase consists of a metal-binding domain and a helicase-binding domain. It undoes the DNA and RNA and facilitates their opening in the 5′−3′ direction. Despite many limitations of this enzyme target, it is considered to be one of the efficient targets against SARS-CoV-2 [23].

3.3 Virus lifecycle

3.3.1 Life process of severe acute respiratory syndrome 2

The virus can be transmitted between humans and animals like a bat, human to human spread by direct contact with the infected person/exposure to cough, sneeze, or respiratory droplets, which creep into the human body through inhalation by nose/mouth [7].

3.3.1.1 Attachment and entry

Entry begins with the strong interaction of viral S proteins with angiotensin-converting enzyme 2 (ACE2) receptors that are located in various organs like the heart, lungs, and kidneys in humans. Later, the fusion of the virus and cell membrane takes place and results in the liberation of the virus into the cytoplasm [24].

3.3.1.2 Replication and transcription

The released viral RNA is now prepared for translation and results in the formation of open reading frames (ORFs) that encrypt various structural and NSPs.

Upon expression of nonstructural polyproteins encoding gene, two polyproteins, namely, pp1a and pp1ab, are formed, which are further proceeded for clinging by CoV proteases (3CL protease and papain-like protease). These are further transformed into NSPs. Later, the replicase−transcriptase complex (RTC) is established, comprising RNA-dependent RNA polymerase (RdRp) and helicase subunit. This complex transcribes the endogenic viral genome template to harmful sense progeny and subgenomic intermediary products, which are later modified to positive sense mRNAs by RdRp. The subgenomic strand encrypted proteins are modified as structural proteins M, S, and E that are enclosed and moved to an intermediary compartment of the endoplasmic reticulum and Golgi apparatus [25,26].

3.3.1.3 Assembly and release

The membrane protein directs the interactivity of the proteins necessary for the coronavirus assembly. It also ties up with the nucleocapsid. Finally, packaging signals are initiated, fulfilling the formation of complete virions. Later, these are liberated by exocytosis (Fig. 3.2) [27].

3.4 Currently available viral targeting drug candidates at various stages of life cycle

These are generally classified into five types, as shown in Table 3.2.

3.5 Different drug repurposing approaches

3.5.1 Target approach

This method of drug repurposing is referred to as the virtual screening approach. Here, various drug candidates that are already established in the libraries are

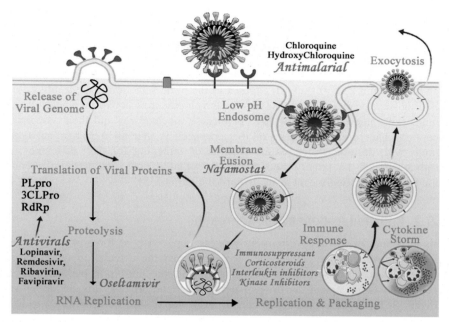

FIGURE 3.2

Drug targets for the virus lifecycle.

Table 3.2 List of viral targeting drug candidates.

Inhibitor type	Mechanism	Example
Entry inhibitors	Virus entry can be inhibited by neutralizing antibodies, proteins/small-molecule compounds.	Chloroquine and hydroxyl chloroquine, etc., [28]
Genome replication inhibitors	Interferes with the RNA polymerase and eludes proofreading, leading to a decline in viral genome production.	Favipiravir [29]
Helicase inhibitors	Unwinds activity of central dogma in the virus life cycle.	Bananins, iodobanani, etc., [30]
Protease inhibitors	Terminates the action of proteases to put an end to virus multiplication.	Lopinavir/ritonavir [31]
Release inhibitors	Inhibits the liberation of new progeny virions from the host cell.	Oseltamivir [32]

utilized for the study purpose. The drug molecules selection criteria depend upon the particular interest and the disease association. This approach is a successful process for recognizing active lead compounds with enhanced properties against the specific disease. The most popular target-based approach in use is high-throughput screening (HTS) in which a huge amount of drug molecules is effectively screened. Other approaches in this category are extra precision (EP) and standard precision (SP) [33,34]. This target-based approach concentrates mainly on the structure of the chemical compounds based on which the lead molecule is identified within a short time. In contrast to the blind selection of drug molecules, this drug repurposing approach offers the development of active therapeutic compounds with ameliorated knowledge of disease mechanisms and their associated receptors. The prime advantage of this approach is that all the chemical compounds with definite structures are capable of screening for their efficiency. A disadvantage of the target-based approach is that it does not permit the compounds of unspecified mechanisms toward the target [35].

3.5.2 Knowledge-dependent approach

This approach of the drug discovery process makes use of foregoing accessible organic information for pathology mechanisms and the association of its receptors. Here, drug-related aspects like mechanisms, the inhibition pathway, target receptors associated with the drug or new entity, and adverse drug reactions are utilized [36]. All these data are obtained to develop a specific drug target to heal a disease. This approach is commonly used to identify adverse drug reactions about the new chemical entities released in the market for rare diseases [37]. It is also useful in identifying new targets for existing drugs, their interactions with the new molecules, and studying the various pathways by which the therapeutic effect is obtained. Apart from all this, information on biomarkers used to identify a specific disease is also obtained. Most importantly, this type of drug repurposing uses the existing information on drugs to predict new therapeutic outcomes to improve accuracy [38].

3.5.3 Molecular docking-based approach

Molecular docking methods are considered to be efficient tools for drug repurposing strategies to recognize new drug molecules. This process mainly works on providing the information for the ability of the drug targets to bind to specific receptors or proteins. By this association between the receptor and drug, we can identify the binding orientation energy of a new molecule toward its target and drug—receptor interactions, which helps understand the disease mechanisms. A drug candidate to be screened by the molecular approach has to satisfy the two components: the searching algorithm and the scoring function [39]. These searching algorithms in molecular docking can identify the strength of binding capacity between the new drug candidate and its receptor. It simply recognizes the active site of the drug on the protein or receptor that can benefit the therapeutic efficacy [40]. For this purpose, different algorithms, namely, genetic algorithms, systematic search, and Monte Carlo, are used. In all these algorithms, a score is estimated named the scoring function based upon the drug interaction on the active site on the receptor. This score indicates the newly repurposed drug molecule's binding capacity and interaction robustness. Currently, these methods are used to estimate protein–protein interactions, protein–drug interactions, etc. The broad categories of the scoring function include knowledge-based, physical-based, and empirical-based. Knowledge-based scoring function utilizes the statistical approach to analyze the binding strength between the drug (protein) and its receptor. Here, the information collected is mainly structures of new molecules, receptors, and the forces of attraction between the protein and the ligand. Therefore, this approach benefits the identification of protein and ligand structural and chemical nature. In the physical-based scoring, information regarding the van der waals and electrostatic forces of attraction are estimated between the protein and ligand. Finally, in the empirical approach, available experimental information is utilized to discuss the levels of interaction between the drugs and ligands [41].

3.5.4 Machine learning approaches

Machine learning methods have gained a spotlight in drug repurposing strategies for their efficient prediction of protein–ligand interactions. These methods provide accurate computation concerning energy changes (entropy) and solute interactions in the solvent on protein–ligand complexes. These methods are widely two types, namely, supervised learning and unsupervised learning. In the supervised learning approaches, information is provided in the form of input, and the primary purpose is to obtain the findings in the form of output in an unbiased manner. This requires data categorization to get the output. In unsupervised machine learning, there are no classifications or divisions in the input provided. The most common benefits of unsupervised learning are data gathering and linearity [42]. While in drug repurposing approaches, supervised machine learning was most routinely used, however, unsupervised approaches as well take part in deciding roles. The sequential tasks associated with the machine learning approaches are data collection, preprocessing of collected data, building datasets, model training, and evolution [43]. Different algorithms employed in machine learning approaches against different diseases in drug repurposing are support vector machine, random forest, deep learning, and deep neural networks (DNNs). The precision of the machine learning model availed in the study depends on the input quality, the sort of datasets employed, and the development of validation procedures. Suppose a workflow is replicated by the above features against a targeted receptor or protein. In that case, the model is successful in the prognostication of active drug molecules with similar analogous characters [44].

3.5.5 Pathway-based approaches

This strategy uses biochemical reactions and signaling mechanisms between the disease and drug to estimate the association of their therapeutic efficiency. There is enormous attentiveness in the computational investigation of drug datasets to apprehend the discovery of new molecule entities as a therapy for a particular disease. Presently, there is expeditious collection of synthetic data and genetic information on various diseases; hence, this is useful in identifying drug targets and their structural relationship to a specific disease. Some software tools like PANTHER and Network Analyst where the information related to defective genes of disease concerning individuals are stored as a repository. Hence, such repository datasets are utilized in analyzing different defective genes and constructing gene lists with their associations in the different disease outcomes by respective tools like Database for Annotation and Visualization for Integrated Discovery [45,46].

3.5.6 Artificial neuronal network approaches

Artificial neuronal networks (ANNs) are popularly used drug repurposing approaches to unsystematically limit the quantitative structure−activity relationship (QSAR) model. This was developed by Bernard. ANN framework possesses many layers of nodes, so biological neurons are interconnected in humans. Usually, nerve cell consists of a branched structure of dendrites, soma, and long axon. Axons help transmit information in the form of action potential from one neuron to another. Similarly, different layers in the ANN imitate the human neurons in terms of interpretation and analysis of existing data [47]. The first ANN framework, perceptron, comprises one input and output that are interlinked to perform various tasks. This model has several benefits in contrast to the statistical model in terms of linear and nonlinear patterns. By using ANN models, we can also categorize the different properties of molecules that are not possible by QSAR strategies. Therefore, this feature is extensively utilized in drug discovery activities to recognize the compounds with potential therapeutic activity [48].

3.5.7 Deep learning machine approaches

The most popular application in the use of this model was DNN that is considered an add-on of ANN. This learning method is useful in studying the extent and scope of big dimensions in addition to unregulated or unstructured information for machine learning. DNN surpassed the other machine learning techniques in drug discovery, especially in estimating the biological nature of drug molecules, pharmacokinetics, and toxicity profiles. Therefore, this model gained a special spotlight in drug discovery and biomarker identification. It is also useful as a drug repositioning approach in the process of drug development and drug discovery. This model avails multiple neuronal layers to separate the high amount of data given in the form of input. It minimizes the modeling error that might arise due to the close relation of function with the dataset. The main features of this model are it can convert networks into small sizes, and it can synchronize the neural dropout. DNN is also fitted with feature extraction, an essential component in pattern recognition to reduce the dimensionality for quantifying high-quality information or data [49,50].

Convolutional neuronal network (CNN) is the most familiar machine learning model of deep learning-based approaches that are used in image analysis. It is established as a successful self-governing method. The architecture of CNN has been made up of an input layer, convolutional layers, and an output layer. These have a characteristic feature of the many-layered perceptron. Apart from image identification, CNN is also used in studies related to the affinity of interactions between ligands and protein molecules. This is also used to study the three-dimensional structure of proteins in the form of 3D images. It is also a well-established model to predict X-ray and CT scan abnormalities in hospital settings that can aid in diagnostic purposes. Hence, these deep learning approaches take part in the drug repurposing methods besides structure identification of proteins [51–53].

3.5.8 Network modeling approach

The network model approach is a widely used procedure in drug repurposing. It can find new drug targets concerning different proteins. This model extensively analyzes various interactions like drug–protein, drug–disease, and protein–protein. All the gathered information is predicted by using different databases to estimate an outcome. Usually, a network refers to different structures representing various variables by nodes, and their relationships are determined by edges. In this model, nodes constitute either a drug/disease, and the edges mean the relation. Generally, the network models are classified as cluster approaches and propagation approaches. Here, cluster-based networks are used to determine the relationship between drug targets and proteins, whereas the propagation approach helps transfer data from one node to another [54,55].

3.5.8.1 Autoencoder approaches

Autoencoder (AE)-based approaches are utilized in the areas of unsupervised machine learning programs. The AE essentially comprises of encoder, decoder, and distance function. Encoder flattens the high-dimensional information into lower-dimensional data, and decoder reframes the information provided by the encoder in a regulated fashion. Distance function usually estimates the data lost in the meanwhile of the generation of input and output. AE produces output features similar to the input given [56]. The most beneficial purpose of AE is it can analyze the unstructured data given in the form of input. AE is employed in the spotting of drug likeliness, drug–disease association, and target-based molecule production toward the specific disease [57,58]. Hence, autoencoding approaches are used to discover the repurposed molecules in times of viral outbreaks.

3.5.8.2 Text mining approaches

Text mining also referred to as text analytics is a similar approach to AI method that utilizes natural language processing. Here, the data or information in an unorganized manner is made into an organized or structured format using different datasets. The output given by this method is embodied in various databases and helps to facilitate the predictive analysis of compounds. Evidence from medical and pharmaceutical fields may contain a vast range of data regarding drugs, diseases, and others [59]. This evidence-based information is drawn as a likely origin of the manifestations of present drug molecules available in the market [60]. For example, in the present corona outbreak, there cannot be available exact evidence related to virus pathogenesis and its therapy because of undefined clinical symptoms and high spread among the population. Therefore, in such a situation, AI-based text mining is utilized to get exact evidence related to the drug—disease relationships and also the viral biomarkers. Biological data available from different datasets are compared using text mining approaches to summarize the relevant information in Table 3.3 [55]. Steps involved in text mining to gather this information precisely include data extraction/information retrieval, biological name identification, and biological knowledge uncovering. All these different steps help understand disease and drug relationships to the specific illness. A summary of different drug repurposing methods is shown in Fig. 3.3.

Table 3.3 Representation of different datasets used for drug repurposing methods.

Category	Databases
Structure-related information	Drug bank
	PubChem
	ChemSpider
Drug target identification	Drug bank
	Therapeutic targets database
	Manually annotated (MATADOR)
Pathway information	Drug Map Central
	NCI database
	Bio Carta
Protein interaction identification	Biological general repository for interaction datasets.
	Protein interacting database.
Genetic information	Online Mendelian Inheritance in man
	Sequence read achieved.
Drug omics data	Encyclopedia

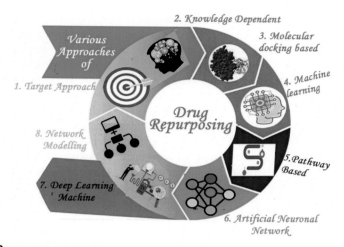

FIGURE 3.3

An overview of drug repurposing approaches.

3.6 Artificial intelligence algorithms for drug repurposing

The most popularly used AI algorithms in drug repositioning are deep learning models and graphical representation for learning purposes. Deep learning models are a subdiscipline of machine learning which mentions the pattern of information identification used by different linear and nonlinear layers organized in an orderly manner in a neuronal network or model [61]. Among the deep learning model, ANNs had gained special interest concerning drug repositioning. This type of programming and data processing system is similar to living neurons of animal intellect. It had also been referred to as connectionist systems. It consists of nodes called neurons that mimic nerve cells of the human cerebrum. Every link looks the same in the way of myoneural junctions of the human cerebrum that process information to the next nerve cells or neurons. In a simple way, a signal is received and transmitted to the next level of nerve cells that make all the neuronal systems activate.

Another network referred to as a fully connected neuronal network (FNN) is an artificial network model that is interconnected by the different input layers to provide the target outputs. This model is usually in use for the information represented as vectors. It can differentiate the drugs into different categories depending on the transcriptome or messenger RNA molecules. In contrast to conventional network models, FNN provides better performance features for the identification of new drug candidates and their respective target sites. Sometimes the output may be in form of images in such a scenario weights associated with the FNN may become too large to process the required output. Therefore, in such a case convolutional models are used. These possess a small filter in between its layers to minimize the pixel or image size. This model is being used to assess the chemical structures of different drug targets [61]. This can be explained with an example of the atom net model where their structural organization estimates a binding affinity or energy of therapeutic candidates. Biologics sequencing is another domain where AI is used as a drug repurposing approach. Here, previous models like FNN and CNN are of limited use. Hence in such sequencing areas, recurrent neural networks are utilized. These models act like drug libraries for the new drug candidates. A hybrid model named molecular transformer—drug target interaction developed by Beck and his colleagues is used to prognosticate the therapeutic efficacy of antiviral drugs in the treatment of SARS-CoV-2. This model recognized several antiviral drugs like atazanavir, efavirenz, and remdesivir as some therapeutic drug candidates for novel coronavirus [62].

A definite procedure of drug repurposing is through a type of graphical representation for learning purposes. This uses network medicine, which builds medical graphs with information on different medical concepts like drugs, disease, receptors, and their existing association with the COVID-19 experimental drugs. The graphs developed by this procedure are termed link estimation graphs with vectors like edges and nodes [63–65]. This peculiar characteristic identifies the degree of association between a given disease and a new therapeutic drug candidate. The biggest confront to this graph prediction model is expandability. Current machine learning approaches like PyTorch avail the use of regular structured information of limited use. However, in the real world, the data produced by medical knowledge are of large scale. Therefore, to cope up with this problem, a new performance system called GraphVite had been developed by Zhu and his colleagues. It can be used as an effective way of drug repurposing since it can process millions of nodes and edges.

3.7 Computational intelligence-based approaches to identify therapeutic candidates for repurposing against coronavirus

Computational drug repurposing methods that are applied for COVID-19 are broadly classified into the following [66]:

1. Network-based models,
2. Structure-based approaches,
3. AI approaches.

3.7.1 Network-based model

These approaches have the potential to integrate numerous data sources; hence, these are widely used in drug repositioning. In this model, the network node signifies drugs, diseases, or gene products, and the edges represent the relationship between the nodes. The consequential pattern helps to figure out the structure-guided pharmaceutical and diagnostic research process with the ability to recognize possible biological targets [66]. The most appropriate network models that are used for COVID-19 are summarized in Table 3.4.

3.7.2 Structure-based approaches

This involves the virtual screening that aids in recognizing small chemical molecules that can bind to known molecular targets. Millions of compounds can be screened within a short period. This can minimize the expenses associated with finding hits to develop new drugs along with new targets for existing drugs [66]. The most appropriate structure-based approaches that are used for COVID-19 drug repurposing are summarized in Table 3.5.

These structure-based approaches also tested the therapies of two antimalarial drugs (chloroquine and hydroxychloroquine) and two anti-AIDS drugs (lopinavir and ritonavir) against COVID-19. They were assumed to bind to many proteins present in the virus envelope. However, no benefit was observed with the anti-AIDS drugs, and the use of hydroxychloroquine and chloroquine is still debated [66].

3.7.3 Artificial intelligence approaches

AI approaches are effective methods to identify marketed drugs with potential antiviral activity against COVID-19. It can help to screen large number of compounds with integrated datasets with the intent to find new drugs for coronavirus. In a paper published by Ke et al., AI technology had been applied to recognize the marketed drugs that had efficacy to treat COVID-19 [67]. This study used two AI dataset models. One AI dataset had the already existing drug compounds with proven activity against COVID-19, human immunodeficiency virus (HIV) and influenza virus, and the other dataset consists of known 3C-like protease inhibitors. The AI model 1 dataset identified 22 drugs with potential antiviral effects. Among these, two promising drugs, clofazimine and gemcitabine, were identified with antiviral activity. On the other hand, AI model dataset 2 identified two molecules, namely, celecoxib and tolcapone with antiviral effects. Therefore, after a few steps of AI learning and prediction techniques old and existing drugs with antiviral activity were identified [67]. Thus, AI plays a crucial part in spotting existing drugs in the pharmaceutical market for the treatment of coronavirus.

Apart from target identification, it can also be implemented in other areas like data storage in clinical settings, early identification and tracking of suspected cases, accurate diagnosis, and helping implement social and preventive measures. Therefore, among the various subgroups of AI, machine learning models and deep learning models help effectively fight against COVID-19 efficiently [68]. A type of deep learning model called CNN is used widely to identify X-ray images of a person with normal pneumonia and COVID-19 pneumonia symptoms. So,

Table 3.4 Drugs repurposed using network-based model.

Algorithm	Method	Network	Description	Possible repurposing drugs
RNSC	Clustering	Protein–protein interaction (PPI)	Helps to identify protein clusters on the PPI network	Tamoxifen, fulvestrant, geldanamycin, loperamide, raloxifene, tanespimycin, alvespimycin
RRW	Clustering	PPI	Effective network clustering method that helps to recognize protein clusters on PPI network	Some complex proteins
ClusterONE	Clustering	PPI	A global network algorithm helps to recognize node clusters in a network	Some complex proteins
ClusterONE	Clustering	Drug–protein–disease	This is a variant of ClusterONE algorithm to cluster nodes on a heterogeneous network	Iloperidone
PRINCE	Propagation	Disease–gene	Helps to identify disease–gene relationship	Some disease–gene relationships
DrugNet	Propagation	Disease–drug–protein	Aids to predict diverse propagation strategies in different subnets	Methotrexate, gabapentin

RNSC, *restricted neighborhood search clustering;* RRW, *repeated random walk;* PPI, *protein–protein interaction.*

Table 3.5 Drugs repurposed using structure-based approaches.

Method	Dataset	Target	Possible repurposing drugs
Direct docking + ensemble docking	FDA-approved drug database + SWEETLEAD database	Main protease (PDB: 6LU7)	Indinavir, ivermectin, cephalosporin-derivatives, neomycin, etc.
Docking + MD simulations	FDA-approved antiviral drugs	Main protease (PDB: 6Y2F)	Lopinavir, ritonavir, tipranavir
Docking	FDA-approved drugs	Main protease (PDB: 6M03)	Glecaprevir, maraviroc
Combined docking	Compounds extracted from ChEMBL database	Main protease (PDB: 6Y2G)	Nearly 64 potential drug targets are identified

this can help to provide an accurate diagnosis for a treatment plan. The procedures that can fall under this category include the Xception and RasNet work which possess diagnostic accuracy of 99% and 91%, respectively. Others, like computed tomography scans, are also utilized in accurate disease identification in contrast to chest X-rays. All these methods provide quick and precise perspectives to recognize the people infected with SARS-CoV-2 and provide them with accurate quarantine measures and treatment strategies. Therefore, present attempts in AI enhance their benefaction for drug development and discovery against the novel coronavirus [69,70]. In the context of drug repurposing, both machine learning and DNN models are being in the application. An online-based deep learning platform named DeepMind helps identify the structural proteins associated with novel coronavirus that can be used for drug screening studies as targets. Drug discovery company Insilico Medicine helps identify the NSP protease structure, an enzyme responsible for viral genome integration into the host genome, by use of drug repurposing. Other than structural protein recognition, these machine learning and AI approaches are also used to outline and bring about new learning algorithms which can be helpful in drug discovery and design approaches. Such algorithms developed by drug repurposing are used in the virtual screening of active molecules or drug targets which are effective toward COVID-19. Spotting a new therapeutic molecule from the scrape is time being and lengthy, therefore not possible to bring it off promptly during pandemics. Hence, repurposing approaches minimize the adverse effects of risk association, toxicity profile, development time for the new molecule candidate, and expenses. The best example of this is the use of the antiviral drug atazanavir, which had been repurposed for COVID-19. Another drug named baricitinib used as a treatment option for arthritis is also used for COVID-19 [71]. This was identified by the UK-based AI company. Viral databases using the drug repurposing approaches of machine learning models had identified the drugs like ribavirin, cyclosporine, and tocilizumab as symptomatic therapy in the treatment of novel coronavirus [67,72]. Drugs like chloroquine, tetracycline, and captopril which target the inhibition of viral entry through ACE2 are also in use for COVID-19. Many others like lopinavir, elbasvir whose target is protease enzyme, arbidol, and chloroquine that targets the spike proteins of COVID-19 are also examples of drug candidates identified by the repurposing approaches in COVID-19 management.

3.8 Challenges in drug repurposing

In the past, medicine had faced several challenges on how to interchange and apply the research findings of new drugs and technologies in the medical field. Despite these drug repurposing applications, existing challenges remain ambiguous. As a part of drug repurposing, animal models cannot provide the same environment with biological conditions as that of the human body. Repurposed drugs also had a greater chance of drug optimization to a particular disease, target, and receptor. During the clinical trials, there might be decreased statistical significance due to small patients' availability and fewer and unclear endpoints [73]. This is well explained with hydroxychloroquine which proves an effective anti-SARS-CoV-2 candidate in animal models; however, it fails to prove its therapeutic efficacy in the preclinical and clinical trials [74−76]. Reproducible animal models for COVID-19 also resulted in a failure in terms of clinical outcomes. For the clinical trials, sometimes patients mildly infected with COVID-19 are recruited. In such a case, there is a need for high-performance tools and analyses to estimate whether the outcome obtained is for a drug or due to placebo. The presence of heterogeneous populations in the trial can affect the drug outcomes because of variations in their genetics. Likely, factors that contribute have a special focus in the clinical trials and are drug candidates targeting incorrect receptor sites, wrong assumptions of pathophysiology mechanisms associated with the SARS-CoV-2 virus; usage of drugs that possess less binding affinity toward the drug−receptor; some drugs mediate at an early stage of disease in mild conditions rather than at the appropriate time; lack of definite biomarkers for the disease identification and drug outcomes, animal models used to estimate the toxicity profiles might have the chances of poor predictive outcomes and efficacy, as there is an urgency to identify the viral mechanisms in host within a limited time; therefore, this can affect the active monitoring of clinical and biological activities of new therapeutic drug candidates. Also, the need for the production of effective therapeutic compounds having beneficial outcomes for *in vitro* and *in vivo* activities might cause the failure of repurposed drugs during the preclinical and clinical stages [74,76].

3.9 Future perspectives of artificial intelligence-informed drug repurposing

The present pharmaceutical market consists of many approved drugs for clinical use. Hence, the selection of appropriate drugs for a particular disease sometimes can be uncertain. In this regard, AI plays a crucial role to clear this uncertainty and proceed with appropriate drug candidates as an option [72]. Hence, AI is considered to be encouraging technology to hasten the drug repurposing for human diseases, particularly in the times of the COVID-19 pandemic outbreak. There is access to information such as clinical, biological, and new AI techniques capable to make avail of information in large datasets for scientific and public domains improving health care [8]. Healthcare professionals like physicians, computer scientists, and statisticians are in a way to adopt AI-based features in the real world for rapid development in the therapeutic area. AI-based algorithms combined with big data had the impact of considerably ameliorating efficacious drug repurposing and assisting in decision-making of drugs during critical needs like COVID-19 [72]. Despite these advantages, challenges like information heterogeneity and lack of sufficient data in clinical settings could affect the development of AI models. To minimize and overcome this, AI-based personalized drug repurposing plays a crucial role. Hence, AI models could hold enormous genomic information to establish the connection between the genetic variants of COVID-19. Therefore, this offers a special chance for personalized treatment in COVID-19 patients. Further, we hope to succeed AI models in drug repurposing domains to be precise in terms of treatment endpoints, source of information, and identification of working mechanisms with vigorous needs.

3.10 Conclusion

The SARS-CoV-2 pandemic upsurge quickly over the globe and has appeared as an exceptional emergency for public health and well-being. To challenge this, preventive measures are needed that greatly reduce the spread of the infection among the public. Despite these preventive efforts, there is a desperate need for medicine or vaccines to fight COVID-19. As the development of new drug entities is time taking, an AI-based drug repurposing approach is an effective way to cope with the current pandemic situation. Because of this, AI models have become an encouraging tool for accelerated drug repurposing, especially in a pandemic situation like COVID-19. Integrating AI techniques with open access data such as repositories is in the high mandate. AI models integrated with big data can have an impact to enhance the effectiveness of repurposed drugs and aids in clinical decision-making. Hence, we can expect in the coming days that AI-based repurposed drugs to be accurate in terms of outcomes and also for effective disease management.

References

[1] H. Li, S.M. Liu, X.H. Yu, S.L. Tang, C.K. Tang, Coronavirus disease 2019 (COVID-19): current status and future perspectives, Int. J. Antimicrob. Agents 55 (2020). Available from: https://doi.org/10.1016/j.ijantimicag.2020.105951.

[2] X. Peng, X. Xu, Y. Li, L. Cheng, X. Zhou, B. Ren, Transmission routes of 2019-nCoV and controls in dental practice, Int. J. Oral. Sci. 12 (2020). Available from: https://doi.org/10.1038/s41368-020-0075-9.

[3] J. Harcourt, A. Tamin, X. Lu, S. Kamili, S.K. Sakthivel, J. Murray, et al., Severe acute respiratory syndrome coronavirus 2 from patient with coronavirus disease, United States, Emerg. Infect. Dis. 26 (2020) 1266−1273. Available from: https://doi.org/10.3201/EID2606.200516.

[4] L. Su, X. Ma, H. Yu, Z. Zhang, P. Bian, Y. Han, et al., The different clinical characteristics of corona virus disease cases between children and their families in China—the character of children with COVID-19, Emerg. Microbes Infect. 9 (2020) 707−713. Available from: https://doi.org/10.1080/22221751.2020.1744483.

[5] A.E. Gorbalenya, S.C. Baker, R.S. Baric, R.J. de Groot, C. Drosten, A.A. Gulyaeva, et al., The species Severe acute respiratory syndrome-related coronavirus: classifying 2019-nCoV and naming it SARS-CoV-2, Nat. Microbiol. 5 (2020). Available from: https://doi.org/10.1038/s41564-020-0695-z.

[6] C.C. Lai, T.P. Shih, W.C. Ko, H.J. Tang, P.R. Hsueh, Severe acute respiratory syndrome coronavirus 2 (SARS-CoV-2) and coronavirus disease-2019 (COVID-19): the epidemic and the challenges, Int. J. Antimicrob. Agents 55 (2020). Available from: https://doi.org/10.1016/j.ijantimicag.2020.105924.

[7] M.A. Shereen, S. Khan, A. Kazmi, N. Bashir, R. Siddique, COVID-19 infection: origin, transmission, and characteristics of human coronaviruses, J. Adv. Res. 24 (2020). Available from: https://doi.org/10.1016/j.jare.2020.03.005.

[8] R. Vaishya, M. Javaid, I.H. Khan, A. Haleem, Artificial intelligence (AI) applications for COVID-19 pandemic, Diabetes Metab. Syndr. Clin. Res. Rev. 14 (2020). Available from: https://doi.org/10.1016/j.dsx.2020.04.012.

[9] A. Alimadadi, S. Aryal, I. Manandhar, P.B. Munroe, B. Joe, X. Cheng, Artificial intelligence and machine learning to fight covid-19, Physiol. Genomics 52 (2020). Available from: https://doi.org/10.1152/physiolgenomics.00029.2020.

[10] R.J. Khan, R.K. Jha, G.M. Amera, M. Jain, E. Singh, A. Pathak, et al., Targeting SARS-CoV-2: a systematic drug repurposing approach to identify promising inhibitors against 3C-like proteinase and 2′-O-ribose methyltransferase, J. Biomol. Struct. Dyn. (2020). Available from: https://doi.org/10.1080/07391102.2020.1753577.

[11] M. Pal, G. Berhanu, C. Desalegn, V. Kandi, Severe acute respiratory syndrome coronavirus-2 (SARS-CoV-2): an update, Cureus (2020). Available from: https://doi.org/10.7759/cureus.7423.

[12] S. Belouzard, J.K. Millet, B.N. Licitra, G.R. Whittaker, Mechanisms of coronavirus cell entry mediated by the viral spike protein, Viruses 4 (2012). Available from: https://doi.org/10.3390/v4061011.

[13] J. Shang, Y. Wan, C. Luo, G. Ye, Q. Geng, A. Auerbach, et al., Cell entry mechanisms of SARS-CoV-2, Proc. Natl Acad. Sci. U. S. A. 117 (2020). Available from: https://doi.org/10.1073/pnas.2003138117.

[14] H. Zhang, J.M. Penninger, Y. Li, N. Zhong, A.S. Slutsky, Angiotensin-converting enzyme 2 (ACE2) as a SARS-CoV-2 receptor: molecular mechanisms and potential therapeutic target, Intensive Care Med. 46 (2020). Available from: https://doi.org/10.1007/s00134-020-05985-9.

[15] Y. Liang, M.L. Wang, C.S. Chien, A.A. Yarmishyn, Y.P. Yang, W.Y. Lai, et al., Highlight of immune pathogenic response and hematopathologic effect in

SARS-CoV, MERS-CoV, and SARS-Cov-2 infection, Front. Immunol. 11 (2020). Available from: https://doi.org/10.3389/fimmu.2020.01022.

[16] J.M. Sanders, M.L. Monogue, T.Z. Jodlowski, J.B. Cutrell, Pharmacologic treatments for coronavirus disease 2019 (COVID-19): a review, JAMA—J. Am. Med. Assoc. 323 (2020). Available from: https://doi.org/10.1001/jama.2020.6019.

[17] C.K. Chang, M.H. Hou, C.F. Chang, C.D. Hsiao, T.H. Huang, The SARS coronavirus nucleocapsid protein—forms and functions, Antivir. Res. 103 (2014). Available from: https://doi.org/10.1016/j.antiviral.2013.12.009.

[18] M. Surjit, S.K. Lal, The SARS-CoV nucleocapsid protein: a protein with multifarious activities, Infect. Genet. Evol. 8 (2008) 397—405. Available from: https://doi.org/10.1016/j.meegid.2007.07.004.

[19] D. Schoeman, B.C. Fielding, Coronavirus envelope protein: current knowledge, Virol. J. 16 (2019). Available from: https://doi.org/10.1186/s12985-019-1182-0.

[20] M. Prajapat, P. Sarma, N. Shekhar, P. Avti, S. Sinha, H. Kaur, et al., Drug targets for corona virus: a systematic review, Indian J. Pharmacol. 52 (2020). Available from: https://doi.org/10.4103/ijp.IJP_115_20.

[21] X. Xue, H. Yu, H. Yang, F. Xue, Z. Wu, W. Shen, et al., Structures of two coronavirus main proteases: implications for substrate binding and antiviral drug design, J. Virol. 82 (2008). Available from: https://doi.org/10.1128/jvi.02114-07.

[22] Y. Gao, L. Yan, Y. Huang, F. Liu, Y. Zhao, L. Cao, et al., Structure of the RNA-dependent RNA polymerase from COVID-19 virus, Science 368 (2020). Available from: https://doi.org/10.1126/science.abb7498.

[23] A.O. Adedeji, B. Marchand, A.J.W. Te Velthuis, E.J. Snijder, S. Weiss, R.L. Eoff, et al., Mechanism of nucleic acid unwinding by SARS-CoV helicase, PLoS One 7 (2012). Available from: https://doi.org/10.1371/journal.pone.0036521.

[24] A.R. Fehr, S. Perlman, Coronaviruses: an overview of their replication and pathogenesis, Coronaviruses Meth. Protoc (2015). Available from: https://doi.org/10.1007/978-1-4939-2438-7_1.

[25] Y. Indwiani Astuti, Diabetes & metabolic syndrome: clinical research & reviews severe acute respiratory syndrome coronavirus 2 (SARS-CoV-2): an overview of viral structure and host response, Diabetes Metab. Syndr. Clin. Res. Rev. 14 (2020).

[26] P.S. Masters, The molecular biology of coronaviruses, Adv. Virus Res. 65 (2006). Available from: https://doi.org/10.1016/S0065-3527(06)66005-3.

[27] R. Poduri, G. Joshi, G. Jagadeesh, Drugs targeting various stages of the SARS-CoV-2 life cycle: exploring promising drugs for the treatment of Covid-19, Cell Signal. 74 (2020). Available from: https://doi.org/10.1016/j.cellsig.2020.109721.

[28] R. Singh, V. Vijayan, Chloroquine: a potential drug in the COVID-19 scenario, Trans. Indian Natl Acad. Eng. 5 (2020). Available from: https://doi.org/10.1007/s41403-020-00114-w.

[29] T.U. Singh, S. Parida, M.C. Lingaraju, M. Kesavan, D. Kumar, R.K. Singh, Drug repurposing approach to fight COVID-19, Pharmacol. Rep. 72 (2020). Available from: https://doi.org/10.1007/s43440-020-00155-6.

[30] X. Zheng, L. Li, Potential therapeutic options for COVID-19, Infect. Microbes Dis. 2 (2020). Available from: https://doi.org/10.1097/im9.0000000000000033.

[31] Y. Song, W. Peng, D. Tang, Y. Dai, Protease inhibitor use in COVID-19, SN Compr. Clin. Med. 2 (2020). Available from: https://doi.org/10.1007/s42399-020-00448-0.

[32] S. Chiba, Effect of early oseltamivir on outpatients without hypoxia with suspected COVID-19, Wien. Klin. Wochenschr. (2020). Available from: https://doi.org/10.1007/s00508-020-01780-0.

[33] E. Lionta, G. Spyrou, D. Vassilatis, Z. Cournia, Structure-based virtual screening for drug discovery: principles, applications and recent advances, Curr. Top. Med. Chem. 14 (2014). Available from: https://doi.org/10.2174/1568026614666140929124445.

[34] M.K. Tripathi, P. Sharma, A. Tripathi, P.N. Tripathi, P. Srivastava, A. Seth, et al., Computational exploration and experimental validation to identify a dual inhibitor of cholinesterase and amyloid-beta for the treatment of Alzheimer's disease, J. Comput. Aided Mol. Des. 34 (2020) 983−1002. Available from: https://doi.org/10.1007/s10822-020-00318-w.

[35] S. Joshua Swamidass, Mining small-molecule screens to repurpose drugs, Brief. Bioinform. 12 (2011). Available from: https://doi.org/10.1093/bib/bbr028.

[36] M. Schenone, V. Dančík, B.K. Wagner, P.A. Clemons, Target identification and mechanism of action in chemical biology and drug discovery, Nat. Chem. Biol. 9 (2013). Available from: https://doi.org/10.1038/nchembio.1199.

[37] R.A. Hodos, B.A. Kidd, K. Shameer, B.P. Readhead, J.T. Dudley, In silico methods for drug repurposing and pharmacology, Wiley Interdiscip. Rev. Syst. Biol. Med. 8 (2016) 186−210. Available from: https://doi.org/10.1002/wsbm.1337.

[38] G. Jin, S.T.C. Wong, Toward better drug repositioning: prioritizing and integrating existing methods into efficient pipelines, Drug. Discov. Today 19 (2014). Available from: https://doi.org/10.1016/j.drudis.2013.11.005.

[39] X.Y. Meng, H.X. Zhang, M. Mezei, M. Cui, Molecular docking: a powerful approach for structure-based drug discovery. Current computer-aided drug design, Curr. Comput. Aided Drug. Des. 7 (2011).

[40] L. Pinzi, G. Rastelli, Molecular docking: shifting paradigms in drug discovery, Int. J. Mol. Sci. 20 (2019). Available from: https://doi.org/10.3390/ijms20184331.

[41] G. Sliwoski, S. Kothiwale, J. Meiler, E.W. Lowe, Computational methods in drug discovery, Pharmacol. Rev. 66 (2014). Available from: https://doi.org/10.1124/pr.112.007336.

[42] F. Napolitano, Y. Zhao, V.M. Moreira, R. Tagliaferri, J. Kere, M. D'Amato, et al., Drug repositioning: a machine-learning approach through data integration, J. Cheminform. 5 (2013). Available from: https://doi.org/10.1186/1758-2946-5-30.

[43] Y.C. Lo, S.E. Rensi, W. Torng, R.B. Altman, Machine learning in chemoinformatics and drug discovery, Drug. Discov. Today 23 (2018) 1538−1546. Available from: https://doi.org/10.1016/j.drudis.2018.05.010.

[44] J. Vamathevan, D. Clark, P. Czodrowski, I. Dunham, E. Ferran, G. Lee, et al., Applications of machine learning in drug discovery and development, Nat. Rev. Drug. Discov. 18 (2019). Available from: https://doi.org/10.1038/s41573-019-0024-5.

[45] J. Li, Z. Lu, Pathway-based drug repositioning using causal inference, BMC Bioinform. 14 (2013). Available from: https://doi.org/10.1186/1471-2105-14-S16-S3.

[46] J.K. Yella, S. Yaddanapudi, Y. Wang, A.G. Jegga, Changing trends in computational drug repositioning, Pharmaceuticals 11 (2018). Available from: https://doi.org/10.3390/ph11020057.

[47] D. Dobchev, M. Karelson, Have artificial neural networks met expectations in drug discovery as implemented in QSAR framework, Expert. Opin. Drug. Discov. 11 (2016). Available from: https://doi.org/10.1080/17460441.2016.1186876.

[48] H. Chauhan, J. Bernick, D. Prasad, V. Masand, The role of artificial neural networks on target validation in drug discovery and development, Artif. Neural Netw. Drug. Des. Deliv. Dispos (2016). Available from: https://doi.org/10.1016/B978-0-12-801559-9.00002-8.

[49] M.M. Najafabadi, F. Villanustre, T.M. Khoshgoftaar, N. Seliya, R. Wald, E. Muharemagic, Deep learning applications and challenges in big data analytics, J. Big Data 2 (2015). Available from: https://doi.org/10.1186/s40537-014-0007-7.

[50] S. Ekins, The next era: deep learning in pharmaceutical research, Pharm. Res. 33 (2016). Available from: https://doi.org/10.1007/s11095-016-2029-7.

[51] P. Schneider, W.P. Walters, A.T. Plowright, N. Sieroka, J. Listgarten, R.A. Goodnow, et al., Rethinking drug design in the artificial intelligence era, Nat. Rev. Drug. Discov. 19 (2020). Available from: https://doi.org/10.1038/s41573-019-0050-3.

[52] S. Hu, C. Zhang, P. Chen, P. Gu, J. Zhang, B. Wang, Predicting drug-target interactions from drug structure and protein sequence using novel convolutional neural networks, BMC Bioinform. 20 (2019). Available from: https://doi.org/10.1186/s12859-019-3263-x.

[53] T. Ozturk, M. Talo, E.A. Yildirim, U.B. Baloglu, O. Yildirim, U. Rajendra Acharya, Automated detection of COVID-19 cases using deep neural networks with X-ray images, Comput. Biol. Med. 121 (2020). Available from: https://doi.org/10.1016/j.compbiomed.2020.103792.

[54] Y. Zhou, Y. Hou, J. Shen, Y. Huang, W. Martin, F. Cheng, Network-based drug repurposing for novel coronavirus 2019-nCoV/SARS-CoV-2, Cell Discov. 6 (2020). Available from: https://doi.org/10.1038/s41421-020-0153-3.

[55] F. Cheng, I.A. Kovács, A.L. Barabási, Network-based prediction of drug combinations, Nat. Commun. 10 (2019). Available from: https://doi.org/10.1038/s41467-019-09186-x.

[56] T. Blaschke, M. Olivecrona, O. Engkvist, J. Bajorath, H. Chen, Application of generative autoencoder in de novo molecular design, Mol. Inf. 37 (2018). Available from: https://doi.org/10.1002/minf.201700123.

[57] Y. LeCun, G.H. Yoshua Bengio, Deep learning (2015), Y. LeCun, Y. Bengio, G. Hinton, Nature (2015).

[58] Q. Hu, M. Feng, L. Lai, J. Pei, Prediction of drug-likeness using deep autoencoder neural networks, Front. Genet. 9 (2018). Available from: https://doi.org/10.3389/fgene.2018.00585.

[59] S. Dash, S.K. Shakyawar, M. Sharma, S. Kaushik, Big data in healthcare: management, analysis and future prospects, J. Big Data 6 (2019). Available from: https://doi.org/10.1186/s40537-019-0217-0.

[60] R. Harpaz, A. Callahan, S. Tamang, Y. Low, D. Odgers, S. Finlayson, et al., Text mining for adverse drug events: the promise, challenges, and state of the art, Drug. Saf. 37 (2014). Available from: https://doi.org/10.1007/s40264-014-0218-z.

[61] J.G. Meyer, S. Liu, I.J. Miller, J.J. Coon, A. Gitter, Learning drug functions from chemical structures with convolutional neural networks and random forests, J. Chem. Inf. Model. (2019). Available from: https://doi.org/10.1021/acs.jcim.9b00236.

[62] B.R. Beck, B. Shin, Y. Choi, S. Park, K. Kang, Predicting commercially available antiviral drugs that may act on the novel coronavirus (SARS-CoV-2) through a drug-target interaction deep learning model, Comput. Struct. Biotechnol. J. 18 (2020). Available from: https://doi.org/10.1016/j.csbj.2020.03.025.

[63] J. Tang, M. Qu, M. Wang, M. Zhang, J. Yan, Q. Mei, LINE: Large-scale information network embedding, WWW 2015 - Proc. 24th Int. Conf. World Wide Web, 2015. Available from: https://doi.org/10.1145/2736277.2741093.

[64] A. Bordes, N. Usunier, A. Garcia-Durán, J. Weston, O. Yakhnenko, Translating embeddings for modeling multi-relational data, Adv. Neural Inf. Process. Syst. (2013).

[65] T. Trouillon, J. Welbl, S. Riedel, E. Ciaussier, G. Bouchard, Complex embeddings for simple link prediction, 33rd Int. Conf. Mach. Learn. ICML 2016, vol. 5, 2016.

[66] S. Dotolo, A. Marabotti, A. Facchiano, R. Tagliaferri, A review on drug repurposing applicable to COVID-19, Brief. Bioinform. 22 (2021) 726–741. Available from: https://doi.org/10.1093/bib/bbaa288.

[67] Y.Y. Ke, T.T. Peng, T.K. Yeh, W.Z. Huang, S.E. Chang, S.H. Wu, et al., Artificial intelligence approach fighting COVID-19 with repurposing drugs, Biomed. J. 43 (2020) 355–362. Available from: https://doi.org/10.1016/j.bj.2020.05.001.

[68] Q.V. Pham, D.C. Nguyen, T. Huynh-The, W.J. Hwang, P.N. Pathirana, Artificial intelligence (AI) and big data for coronavirus (COVID-19) pandemic: a survey on the state-of-the-arts, IEEE Access. 8 (2020). Available from: https://doi.org/10.1109/ACCESS.2020.3009328.

[69] D. Ho, Addressing COVID-19 drug development with artificial intelligence, Adv. Intell. Syst. 2 (2020). Available from: https://doi.org/10.1002/aisy.202000070.

[70] M. Tsikala Vafea, E. Atalla, J. Georgakas, F. Shehadeh, E.K. Mylona, M. Kalligeros, et al., Emerging technologies for use in the study, diagnosis, and treatment of patients with COVID-19, Cell Mol. Bioeng. 13 (2020). Available from: https://doi.org/10.1007/s12195-020-00629-w.

[71] J. Stebbing, A. Phelan, I. Griffin, C. Tucker, O. Oechsle, D. Smith, et al., COVID-19: combining antiviral and anti-inflammatory treatments, Lancet Infect. Dis. 20 (2020). Available from: https://doi.org/10.1016/S1473-3099(20)30132-8.

[72] S. Mohanty, A.I. Harun, M. Rashid, M. Mridul, C. Mohanty, S. Swayamsiddha, Application of artificial intelligence in COVID-19 drug repurposing, Diabetes Metab. Syndr. Clin. Res. Rev. 14 (2020). Available from: https://doi.org/10.1016/j.dsx.2020.06.068.

[73] H.B. Mehta, S. Ehrhardt, T.J. Moore, J.B. Segal, G.C. Alexander, Characteristics of registered clinical trials assessing treatments for COVID-19: a cross-sectional analysis, BMJ Open. 10 (2020). Available from: https://doi.org/10.1136/bmjopen-2020-039978.

[74] J. Liu, R. Cao, M. Xu, X. Wang, H. Zhang, H. Hu, et al., Hydroxychloroquine, a less toxic derivative of chloroquine, is effective in inhibiting SARS-CoV-2 infection in vitro, Cell Discov. 6 (2020). Available from: https://doi.org/10.1038/s41421-020-0156-0.

[75] S.J. Park, K.M. Yu, Y. Il Kim, S.M. Kim, E.H. Kim, S.G. Kim, et al., Antiviral efficacies of FDA-approved drugs against SARS-COV-2 infection in ferrets, MBio 11 (2020). Available from: https://doi.org/10.1128/mBio.01114-20.

[76] E.S. Rosenberg, E.M. Dufort, T. Udo, L.A. Wilberschied, J. Kumar, J. Tesoriero, et al., Association of treatment with hydroxychloroquine or azithromycin with in-hospital mortality in patients with COVID-19 in New York State, JAMA—J. Am. Med. Assoc. 323 (2020). Available from: https://doi.org/10.1001/jama.2020.8630.

COVID-19: artificial intelligence solutions, prediction with country cluster analysis, and time-series forecasting

Sreekantha Desai Karanam[1], Rajani Sudhir Kamath[2] and Raja Vittal Rao Kulkarni[2]

[1]NITTE (Deemed to be University), NMAM Institute of Technology (NMAMIT), Nitte, Karnataka, India
[2]CSIBER, Kolhapur, Maharastra, India

4.1 Introduction

The novel coronavirus 2019 (SARS-CoV-2) outbreak started at beginning of December 2019 in China in Wuhan city. This viral infection requires about 5.2 days for development and triggers symptoms such as fever, cough, and flu. This coronavirus-induced disease was named COVID-19, which can damage many tissues and organs. Many infected patients showed the symptoms of pneumonia called novel coronavirus pneumonia (NCP) which quickly advances into a severe respirational problem with higher mortality and the least prognosis. China has adopted unprecedented nationwide interventions from January 23, 2020, when transmission of the virus from person to person was established. These stringent measures on travel, widespread monitoring of suspected cases, and significant quarantining of patients were imposed to contain the spread of the COVID-19 outbreak. Vaccinating billions of people across the world in a short time was a challenge. This pandemic accelerated the research and applications of artificial intelligence (AI) to fight against COVID-19. Healthcare service providers immediately require a decision support system to control the COVID-19 spread and support them in providing appropriate recommendations online. AI techniques that are integrated with big data can significantly increase the efficacy and productivity of drug repurposing and speed up medical decision-making.

4.1.1 Motivation for this study

The pandemic has resulted in several thousand billion dollars of productivity and monetary loss. This loss may be reduced by the discovery of infected patients, tracking of

Artificial Intelligence in Healthcare and COVID-19. DOI: https://doi.org/10.1016/B978-0-323-90531-2.00010-2

affected places, and prediction of evolving hot spots. This pandemic has uncovered innumerable shortages in the quantity and quality of healthcare services and data management. There is an urgent requirement to provide online information to caretakers, doctors, and infected patients. This pandemic has also presented distinctive problems and prospects for ML and AI researchers to collaborate with clinical professionals to quickly implement AI-powered solutions for treating evolving diseases. The cooperation between AI professionals, researchers, and governmental agencies was very much needed to build AI-based tools which are accurate, relevant, and create good impact.

4.1.2 Adverse impacts of COVID-19 outbreak

Kimberly Chriscaden, Communications Officer of WHO, has quoted that human life worldwide was worst affected by the COVID-19 pandemic which lead to public health disasters across the world [1]. The pandemic has highlighted the present inadequate healthcare infrastructure and the low doctor-to-patient ratio, resulting in shortage of ventilators, beds, biosafety measures, insufficient testing capacity, resources, and a surge in patient contacts and queries, deepening the despairs of the healthcare sector. This pandemic has significantly worsened the world economy, financial markets, public health, and food systems. A substantial decline in income, increased unemployment levels, and disruptions in every service, business, and industry across the world. Globally, many countries have misjudged the risks of the pandemic leading to a swift spreading rate and later reacted to the public health crisis that magnified the calamity. International proactive actions were expected to be implemented to save the lives of millions of people and to protect world economic prosperity. Millions of enterprises are struggling for survival. About 50% of 3300 million workers across the world were on the verge of losing their incomes. All national and international supply chains were worst affected by the huge losses to business and industry sectors.

The governments are struggling with new lockdown measures to combat the spread of the virus [2]. The share markets are showing very much downtrends. In the United States, about 8.9% of the workforce was affected by lockdowns as disclosed by International Monetary Fund (IMF). Millions of employees are covered by government-supported job retention policies. The airline, tourism, and hospitality industry are worst affected. The IMF stated that the global economy was reduced by 4.4% in 2020. Only pharmaceutical company shares such as Moderna, Novavax, and AstraZeneca that are producing a vaccine for COVID-19 has shown an upward trend. The International Air Transport Association (IATA) has estimated a loss in airline passenger travel income up to $314 billion. In Europe, the gross domestic product (GDP) of Italy, Spain, and France fell by 17.5%, 19.2%, and 21.3%, respectively. The most recent analysis cautions that COVID-19 has pressed 88 million more people to poor and life-threatening conditions in the current year [3]. Many companies have adopted digital technologies and promoted existing unemployment cases. COVID-19 has highlighted the requirement for efficient and cost-effective healthcare services. Citizens in developing countries are paying more than 500 billion dollars from their pockets for healthcare. These excessive costs are driving economic problems to about 900 million more people and forcing about 90 million people into dangerous poverty status every year. During these pandemic times, about 160 countries have been forced to close down their schools disturbing at least 1.6 billion children and youth from learning.

With the losses of learning and increased school dropouts, the students leading to estimated to loss of $10 trillion in income, which is about 10% of the global GDP. The Internet is indispensable for several services, such as e-payment systems, e-health platforms, and digital cash transfers was extensively used during these pandemic times. COVID-19 has added about 132 million people to the total cases of malnourished across the globe in 2020.

The highlights of COVID-19 adverse impacts are

Healthcare industry

- Challenges in testing, isolation, and management of suspected, infected patients;
- Overburdening of existing medical resources and medical shops;
- The patients suffering from other chronic diseases are not getting enough care;
- Excess load on nurses, medical practitioners, other health workers, these people were exposed to severe health risks and also lost their lives;
- Demand for highly safe and protective equipment; and
- Distraction and shortages in the medical supply chain.

Economic

- The slowdown in the production of essential items and delivery of services;
- Delay in the transportation of products from source to destinations;
- Reduction and slowdown in the growth of domestic and world trade incomes; and
- Less currency flowing in the market.

Society

- The service industry was unable to meet the customer demands;
- Termination of international games and competitions;
- Prevention domestic, global tours, travels, and termination of flights;
- Cancellation of social, spiritual, and celebration events;
- Excessive tension and mental stress on the common man;
- Need for maintaining social distance with friends and family members;
- Shutting of religious places, restaurants, and hotels;
- Shutting of schools, colleges, swimming pools, gymnasiums, learning centers, theaters, sports clubs, and many more;
- Delay or termination of offline classes and examinations and switching to online classes and examinations (Figs. 4.1 and 4.2).

4.1.3 Chapter organization

This chapter is organized into six sections. Section 4.1 introduces the concepts, and Subsection 4.1.1 deals with the motivation for study. Subsection 4.1.2

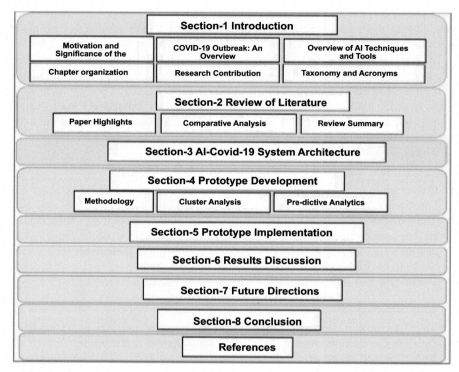

FIGURE 4.1

Chapter organization.

shows the adverse impacts of the COVID-19 outbreak, Subsection 4.1.3 shows the organization of this chapter, and Subsection 4.1.4 presents the table of abbreviations used. The study of the current literature is illustrated in Section 4.2. Section 4.2 is distributed into two subsections: Subsection 4.2.1 describes the papers reviewed from one technology domain. The highlights and findings from each paper are presented in a table. Subsection 4.2.2 deals with research findings and summary. Section 4.3 outlines K-means clustering for COVID-19 country-wise analysis. Section 4.4 discusses the time-series modeling for COVID-19 new cases. Section 4.5 presents the conclusion. The last sections show the references of the research papers surveyed.

4.1.4 Table of acronyms used in this chapter

This chapter has used a set of acronyms for describing the contents of the chapter. Table 4.1 shows all those acronyms for easy reference.

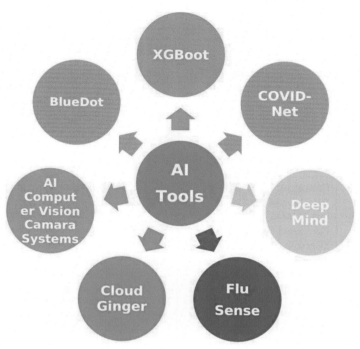

FIGURE 4.2

COVID-19 vaccine companies.

4.2 Review of literature on COVID-19 pandemic

The authors have carefully selected papers from high-impact factor journals from Springer, Science Direct, IEEE, and Elsevier publications. About 44 research papers covering COVID-19-related concepts were studied. This section discusses the interesting points noted in every paper by the way of illustration. The section 4.2.1 and 4.2.2 are included in annexure as advised by reviewers.

A study on the role of image-based AI in treating comorbid patients having COVID-19 symptoms was carried out [4]. The symptoms identified are arrhythmias, coronary thrombosis, myocardial injury, hypoxia, plaque rupture, venous thromboembolism, lung injury, ischemia, encephalitis, and inflammation. Image-powered AI can be utilized to identify the tissues of a COVID-19 patient and categorize the seriousness of the infection. Scrutiny of medical images can considerably reveal the patient's probability of survival. AI-based medical imaging methods are used to speed up diagnoses and decision-making in pandemic cases. A review of the medical lab practices with an AI-based electrocardiogram (ECG) to diagnose COVID-19 ventricular dysfunction was discussed [5]. The patients were treated

Table 4.1 List of abbreviations used in this chapter.

Acronyms	Description	Acronyms	Description
AI	Artificial Intelligence	EMR	Electronic Medical Record
COVID-19	CoronaVirus Disease 2019	CCD	Corona Combat Drone
WHO	World Health Organization	TCCD	Thermal Corona Combat Drone
NCP	Novel Coronavirus Pneumonia	EBT	Elevated Body Temperature
IMF	International Monetary Fund	UAV	Unmanned Arial Vehicle
GDP	Gross Domestic Product	ECG	ElectroCardioGram
VIR	Virtual Interventional Radiologist	MCDA	MultiCriteria Decision Analysis
UCLA	The University of California, Los Angeles	SEIR	Susceptible-Exposed-Infectious-Removed
CT	Computed Tomography	CXR	Chest Radiographs
MRI	Magnetic Resonance Imaging	RMSE	Root Mean Squared Error
MIT	Massachusetts Institute of Technology	AUROC	Area Under the Receiver Operating Characteristic
CNN	Convolutional Neural Network	PROBATE	Prediction Model Risk of Bias Assessment Tool
CHARMS	Critical Appraisal and data extraction for systematic Reviews of prediction Modeling Studies.	PRECISE	Precise Risk Estimation to optimize COVID-19 Care for Infected or Suspected patients in diverse settings
PCF	Partial Correlation Function	ACF	Auto-Correlation Function
SARS	Severe Acute Respiratory Syndrome	ARIMA	Autoregressive Integrated Moving Average
IATA	International Air Transport Association	TCS	Tata Consultancy Services
ML	Machine Learning	DL	Deep Learning

with echocardiography and electrocardiography after 14 days from the day of the COVID-19 positive test. Only one patient out of 27 patients had a normal ventricular function in the past and subsequently progressed into COVID-19 myocarditis. This case study reveals that AI ECG can be used as a diagnostic device for the discovery of dysfunction of the heart in COVID-19 patients.

A database of 3777 patients' computed tomography (CT) images was studied [6]. The AI-based tool was used to distinguish NCP from other regular pneumonia. This AI tool helped physicians and medical imaging experts in rapid screening for patients in critical condition. This AI tool was made accessible worldwide to help clinicians to fight the COVID-19 disease. AI systems are concentrating on finding the unusable lungs where there is a tendency of patients affected with COVID-19 [7]. The authors demonstrated that no appropriate study benchmarked and assessed AI methods used in medical image classification problems of

COVID-19 [8]. This process was assumed as a multicomplex attribute problem. Multicriteria decision analysis (MCDA) was considered an efficient method to solve this complex problem. Solving this problem leads to many future research directions. The authors revealed that only 11 cases applied AI-based approaches in classifying and identifying COVID-19 cases.

A comprehensive analysis of the significance of AI and ML techniques for prediction, diagnosis, projecting, tracing contacts, and drug design for a COVID-19 pandemic was conducted [9]. A practical AI-based method to distinguish COVID-19 infected patients from normal persons based on X-ray images of the chest and CT scan was presented [10]. This method uses a database of X-ray images of the chest and a decision tree algorithm for detecting COVID-19 infection. The precision of COVID-19 detection was measured using the F1 score. The accuracy of results is based on the quality of CT scans and chest X-ray images stored in Open-I and Kaggle open databases. This method discovers COVID-19 cases with 93% accuracy. AI techniques for predicting infection, diagnosis, death rate by tracking contacts and targeted drug delivery were studied [11]. The influence of various types of medical data applied in prognosis, diagnosis, and analysis of pandemic impact was also discussed. AI-based methods are used in developing models to assess the criticality of disease, sequence formation of virions, and outbreak prediction models.

The authors reviewed the implementations of AI solutions in diverse areas such as diagnosis of diseases, treatment, forecasting criticality of a patient, identification of symptoms, processing of X-ray images and chest CT scans [12]. CNN, DL, ML, and evolutionary algorithms are applied for the management of the Covid-19 cases. This paper presented the problems faced and suggested the methods to overcome them. An AI-based diagnostic app called AI4COVID-19 to detect COVID-19 infection was developed [13]. This app captures and transmits 3 seconds of cough audio to a cloud-hosted engine powered with AI and sends back results within 2 minutes. The authors analyzed the unique path of morphological changes in lungs affected by novel coronavirus disease. Transfer learning method was applied to cover the lack of sufficient COVID-19 cough training data. The authors applied multiple-pronged mediators based on AI risk-averse architecture to decrease the wrong diagnosis risk arising from multiple views. The outcome revealed that AI4COVID-19 can differentiate coughs of COVID-19 from different kinds of coughs of non-COVID-19. AI4COVID-19 can be used as a primary diagnostic app anytime, anywhere, and by anyone. An autonomous AI device to evaluate CT scan images to check the possibility of pneumonia infection leading to COVID-19 was presented [14]. This AI device uses only classification and segmentation techniques for COVID-19 detection and saves the detection time of physicians by 30%−40%. This AI device assigns a ranking to each CT image showing the possibility of infection so that the doctors can endorse and separate the infected patients on time. The authors have used 1136 images for training which consists of 723 COVID-19 positives collected from five healthcare centers and achieved a specificity of 0.922 and accuracy of 0.974 during the testing phase.

A deep learning technique was applied to X-ray images for diagnosing COVID-19 infection [15]. The authors evaluated the efficacy of eight CNN pretrained models such as ResNet-34, MobileNet-V2, ResNet-50, AlexNet, SqueezeNet, GoogleNet, VGG-16, and Inception-V3 for separating infected cases from regular cases. These CNN networks have been assessed on open databases of chest X-ray images, and the maximum efficiency was attained using ResNet-34 with 98.33% accuracy. An AI-powered device was developed that would directly estimate the drugs/peptides from the series of COVID-19 positive cases [16]. The efficiency of AI methods for discovering coronavirus infection in chest X-ray radiographs was tested (CXR). The results are compared to the outcome with those of doctors [17]. This AI method was tweaked to differentiate definitive NCP from other regular pneumonia. Fifty-four doctors masked to diagnosing are called to determine the X-ray images in similar test data conditions and allocated at random to accept or not to accept help from the AI device. The performances were compared between the identification efficiency of doctors who worked with help and without AI device help. The kinetics of the pandemic concerning AI and several parameters of diagnosis and therapy was studied [18]. Innovative ideas were developed, experiences were recorded, and plans for containing the pandemic were described. The authors presented the methods for diagnosing COVID-19 and plans for developing AI-enabled solutions for epidemics. Drug repurposing is a technique in which present vaccines are utilized for the treatment of COVID-19 [19]. Drug repurposing was a potential solution because it reduces research and testing overheads and saves time. The authors introduced a road map for utilizing AI solutions to speed up drug repurposing [20]. AI played a significant role in tracing the cluster of cases and predicting the places in which the virus would infect in the future based on analysis of past data. Authors decided that AI solutions were not very successful in fighting against COVID-19 [21].

Artificial intelligence requires huge data, and it is suffering because of the lack of a large database. A deep convolutional neural network framework using two datasets was trained [22]. These datasets are from the GitHub-COVID repository20, COVID-19 positive radiographs from the National Institutes of Health's ChestX-ray14 repository21. Integration of data about people migrated before and after January 23, 2020 and the most updated COVID-19 epidemiologic data into the susceptible-exposed-infectious-removed (SEIR) model was carried out [23]. The training was based on 2003 SARS data, to forecast the pandemic. Authors estimated that China's pandemic conditions should reach a maximum by end of February, exhibiting a slow reduction by end of April 2020. The dynamic SEIR model was efficacious in anticipating the COVID-19 spread sizes and peaks.

The various techniques that can be used on pandemic data were highlighted [24]. This paper classified the present AI techniques in medical image data analysis. An overview of big medical data processing using AI-based solutions to fight against COVID-19 was presented [25]. The authors pointed out the difficulties faced by AI and bigdata community in combating COVID-19. The improved practice of data sharing should be imposed in the metropolitan healthcare sector [26]. AI systems are processing data from smart cities data sources and wearables for precise detection and prediction of patients with COVID-19 infections. The COVID-19 crisis clutched the globe, and the influence of AI on customer behavior is ever increasing [27]. AI-powered augmented reality glasses are used to detect high-temperature levels in the crowd in a few minutes. AI tools enable us to check whether travelers on trains and buses are wearing face masks or not. The research trends on identifying the potential of AI and improving its competencies in the fight against the pandemic are meticulously explored [28]. This paper highlighted 13 sets of issues associated with COVID-19 disease and suggested suitable AI algorithms that can be applied to solve these issues. The authors assessed the efficacy and legitimacy of distributed reports of forecasting models used for identifying COVID-19 infections [29]. COVID-PRECISE stands for Precise Risk Estimation to optimize COVID-19 Care for Infected or Suspected patients in diverse settings. The databases used for the study are Embase and PubMed through arXiv, bioRxiv, medRxiv, and Ovid up to May 5, 2020. This paper has implemented a multivariable COVID-19-based forecasting model.

OECD website (Read for Latest Updates on Corona) stated that at present AI methods and tools are playing a significant part in all aspects of the COVID-19 crisis [30]. AI tools assist in avoiding or reducing the COVID-19 blowout and primary contact tracking. Authors said that AI could be utilized as an initial pandemic alarming system [31]. An AI-powered app, BlueDot, effectively discovered 9 days before the WHO reported the outbreak of Zika virus in Florida4 and also detected COVID-19, alarming the public about the growth of COVID-19. AI-based robots are used for distributing drugs, food to infected patients, disinfecting surfaces, and helping doctors in managing pandemics [32]. AI-driven tools such as smartphone apps are used for self-diagnosis by patients staying at home. AI-powered scanning devices can be used for detecting infected people in public places. AI-based disease prediction tools can be used as primary diagnostic tools to control the rapid growth of the COVID-19 virus spreading.

4.3 K-means clustering for COVID-19 country analysis

4.3.1 Cluster analysis: an overview

This section reports an unsupervised learning approach for identifying NCP-infected countries based on the varieties of COVID-19 infections. NCP situation reports are refereed for this study [33]. These reports provide worldwide details of the present COVID-19 epidemiological conditions, transmission categorizations death counts, present infected cases, and so on. It is observed from the dataset that the highest number of new cases was 862,181 on January 7, 2021. The authors implemented K-means clustering algorithm for the visualization of infected countries. The clusters of these nations were formed based on the parameters such as total cases per million, reproduction rate, median age, total cases, new cases, life expectancy, and total deaths. This work exhibits the performance of the unsupervised learning technique by experimenting with several iterations and clusters. The sum of square errors was considered from each cluster to measure the performance of the K-means clustering.

4.3.2 Dataset selection and preprocessing

The dataset for the present study was retrieved from an open-access data platform named Our Complete COVID-19 [34]. This COVID-19 dataset was maintained by Our World. The authors considered the dataset on hospitalizations, infected cases, deaths, and testing for all countries since January 22, 2020, managed by Johns Hopkins University [35]. These COVID-19 spread data were extracted from government, international, and subnational authorities around the globe—a complete list of databases for each country was publicized and made data publicly available on Johns Hopkins GitHub site [36]. It was observed from the dataset that the highest number of new cases, that is, 862,181, was recorded on January 7, 2021. The authors considered 192 countries' COVID-19 data as recorded on January 7, 2021, for the present study.

The parameters considered for this study are total cases per million, reproduction rate, median age, total cases, new cases, life expectancy, and total deaths. Some values are missing for the countries such as Andorra, Bhutan, Cambodia, Dominica, Fiji, Grenada, Hong Kong, Kosovo, Laos, Liechtenstein, Marshall Islands, Monaco, Saint Vincent, Saint Kitts, Grenadines, Nevis, Samoa, San Marino, Solomon Islands, Timor, Vanuatu, and Vatican. These countries are excluded from the dataset for accuracy purposes. Thus the derived dataset for this study consists of 170 countries' details with six variables. Fig. 4.3 illustrates the partial dataset. Ci = [total cases I, new cases I, total deaths I, total cases per million i, reproduction rate i, median age I, life expectancy i] where $i = 1-170$.

1	location	total_cases	new_cases	total_deaths	total_cases_per_million	reproduction_rate	median_age	life_expectancy
2	Afghanistan	53207	102	2253	1366.793	0.47	18.6	64.83
3	Albania	61705	697	1223	21441.726	1.18	38	78.57
4	Algeria	101382	262	2792	2311.963	0.81	29.1	76.88
5	Andorra	8348	0	84	108043.746	1.11		83.73
6	Angola	17974	110	413	546.883	1.02	16.8	61.15
7	Antigua and Ba	163	0	5	1664.488	0.51	32.1	77.02
8	Argentina	1690006	13835	44122	37393.007	1.24	31.9	76.67
9	Armenia	161054	201	2901	54350.753	0.7	35.7	75.09
10	Australia	28571	25	909	1120.437	0.9	37.9	83.44
11	Austria	374730	2540	6568	41607.079	1.05	44.4	81.54
12	Azerbaijan	222885	685	2845	21982.558	0.56	32.4	73
13	Bahamas	7959	14	175	20239.137	1.07	34.3	73.92
14	Bahrain	94633	349	353	55614.684	1.12	32.4	77.29
15	Bangladesh	519905	1007	7718	3156.882	0.87	27.5	72.59
16	Barbados	743	0	7	2585.508	1.2	39.8	79.19
17	Belarus	206796	1838	1489	21884.747	0.98	40.3	74.79
18	Belgium	658655	2923	19936	56831.477	1.05	41.8	81.63
19	Belize	11152	44	267	28046.808	0.9	25	74.62
20	Benin	3304	0	44	272.535	0.79	18.8	61.77
21	Bhutan	767	12		994.023	1	28.6	71.78
22	Bolivia	168891	1910	9304	14468.481	1.51	25.4	71.51
23	Bosnia and Her	113392	0	4211	34562.144	1	42.5	77.4
24	Botswana	16050	610	48	6825.068	1.01	25.8	69.59
25	Brazil	7961673	87843	200498	37456.225	1.14	33.5	75.88
26	Brunei	173	1	3	395.444	0.07	32.4	75.86
27	Bulgaria	207259	867	8017	29828.113	0.85	44.7	75.05
28	Burkina Faso	7713	150	89	368.985	1.12	17.6	61.58
29	Burundi	885	1	2	74.427	1.12	17.5	61.58
30	Cambodia	386	1		23.088	0.57	25.6	69.82

FIGURE 4.3

Partial dataset for cluster analysis.

4.3.3 Findings from COVID-19 country cluster data analysis

The basic statistical analysis of this dataset is shown in Fig. 4.4. A minimum count of new cases were found in Antigua and Barbados. The maximum number of new cases was found in the United States. The total count of new cases per million is highest in Montenegro whereas least in Tanzania. Total deaths and new cases are also the maxima in the United States. K-means clustering, an unsupervised classification technique was employed for country cluster analysis in this study. The K-means technique derives an assortment of k clusters using a heuristic search initiating with a selection of k randomly selected clusters each of which in the beginning represents a cluster mean [37]. The COVID-19 countries cluster analysis and visualization are simulated in the R system, an open-source data mining platform. The country's COVID-19 records are clustered based on curated six parameters. A single record Ci is represented as a multidimensional data vector and is defined as [38]. K-means clustering needs the analyst to identify the total number of clusters to be designed. The "number of groups against" the "within groups total of squares," which are the parameters that should be minimized during the clustering procedure, was plotted [39]. Fig. 4.5 shows this plot for a different number of clusters. This plot reveals that this amount diminishes to a certain point 5, and afterward, there is not much significant decrease in values.

The clustering depends on estimating the likeness between nations by figuring the separation between each pair. This similarity is assessed based on the mean estimation of the nations in a group. The R function K-means () was applied to get clusters of COVID-19 countries. The analysis was tuned by changing the count of clusters by keeping 15 iterations constant. The experiment is assessed regarding within clusters sum of square errors and between SS and total SS. The ideal value for BSS/TSS, that is, the properties of internal cohesion and external separation should approach 1. Fig. 4.6 shows that BSS/TSS value is 98.6% for five clusters of size 2, 136, 1, 9, and 22.

```
  total_cases          new_cases            total_deaths          total_cases_per_million
Min.   :      163    Min.   :      0     Min.   :      1.0     Min.   :      8.52
1st Qu.:     9691    1st Qu.:     38     1st Qu.:    130.0     1st Qu.: 1209.82
Median :    72385    Median :    356     Median :    923.5     Median : 8635.79
Mean   :   517856    Mean   :   5067     Mean   :  11159.2     Mean   :17004.86
3rd Qu.:   232340    3rd Qu.:   1825     3rd Qu.:   4893.0     3rd Qu.:29821.51
Max.   : 21636431    Max.   : 280855     Max.   : 365324.0     Max.   :81456.93
reproduction_rate    median_age          life_expectancy
Min.   :0.000        Min.   :15.10       Min.   :53.28
1st Qu.:0.850        1st Qu.:21.55       1st Qu.:66.62
Median :1.050        Median :29.80       Median :74.47
Mean   :1.005        Mean   :30.45       Mean   :72.64
3rd Qu.:1.180        3rd Qu.:39.00       3rd Qu.:78.36
Max.   :2.130        Max.   :48.20       Max.   :84.63
```

FIGURE 4.4

Basic statistical summary of COVID-19 infected countries as of January 7, 2021.

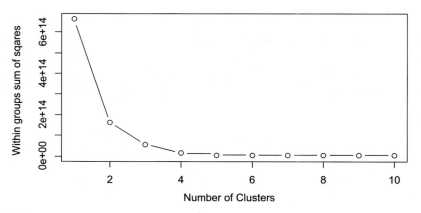

FIGURE 4.5

Number of clusters against within clusters sum of squares.

```
K-means clustering with 5 clusters of sizes 2, 136, 1, 9, 22

Cluster means:
   total_cases    new_cases total_deaths total_cases_per_million
1   9187545.00   52991.0000    175534.000                22501.08
2     75654.68     649.3015      1428.309                13420.53
3  21636431.00  280855.0000    365324.000                65366.34
4   2312726.22   25105.5556     53818.444                34326.03
5    769110.64    7289.9091     22820.909                29378.64
   reproduction_rate median_age life_expectancy
1         0.9800000   30.85000        72.77000
2         0.9814706   29.06544        71.47243
3         1.1700000   38.30000        78.86000
4         1.1177778   39.78889        79.62333
5         1.1027273   34.79091        76.68318

Clustering vector:
  [1] 2 2 2 2 2 4 2 2 2 2 2 2 5 2 2 5 2 2 2 2 2 1 2 2 2 2 2 5 2 2 2 5 2
 [34] 4 2 2 2 2 2 2 2 5 2 2 2 2 2 2 2 2 2 2 2 2 2 4 2 2 2 4 2 2 2 2 2 2
 [67] 2 2 2 2 1 5 5 5 2 5 4 2 2 2 2 2 2 2 2 2 2 2 2 2 2 2 2 2 2 2 2 2 2
[100] 5 2 2 2 5 2 2 2 2 5 2 2 2 2 2 2 5 2 2 2 2 5 5 5 5 2 5 4 2 2 2 2
[133] 2 2 2 2 2 2 2 2 5 2 2 4 2 2 2 5 5 2 2 2 2 2 2 2 2 4 2 5 2 4 3 2 2
[166] 2 2 2 2 2

Within cluster sum of squares by cluster:
[1] 3.009647e+12 1.165868e+12 0.000000e+00 2.514034e+12 2.278533e+12
 (between_SS / total_SS =  98.6 %)
```

FIGURE 4.6

K-means clustering of countries based on COVID-19 cases.

4.3.4 **The results and discussions**

The dataset of 170 countries as of January 7, 2021, was clustered into five groups based on the status of infected cases. The K-means clustering technique was employed for this purpose. The clusters and the participating countries are tabulated in Table 4.2. By analyzing the cluster means, one can relate each group with each of the five classes of countries:

1. Cluster 1 was formed by India and Brazil which are the second highest in the total count of cases, novel cases, and the total number of patients who died.
2. Dataset of 136 countries with the least cluster means for total cases per million, reproduction rate, median age, total cases, new cases, life expectancy, and total deaths.
3. Only the United States is in cluster 3 with the total cases per million, reproduction rate, median age, total cases, new cases, life expectancy, and total deaths.
4. Only nine countries fall under cluster 4 having the highest cluster means for median age and life expectancy.

Table 4.2 Clusters and participating countries.

Cluster 1	Cluster 3	Cluster 4	Cluster 5
India, Brazil	United States	"Afghanistan, Argentina, Denmark, France, Germany, Italy, Russia, Spain, Turkey, United Kingdom"	"Bangladesh, Belgium, Canada, Chile, Czechia, Indonesia, Iran, Iraq, Israel, Mexico, Morocco, Netherlands, Pakistan, Peru, Philippines, Poland, Portugal, Romania, South Africa, Sweden, Switzerland, and Uganda"

"Afghanistan, Albania, Algeria, Angola, Antigua and Barbuda, Armenia, Australia, Austria, Azerbaijan, Bahamas, Bahrain, Barbados, Belarus, Belize, Benin, Bolivia, Bosnia and Herzegovina, Botswana, Brunei, Bulgaria, Burkina Faso, Burundi, Cameroon, Cape Verde, Central African Republic, Chad, China, Colombia, Comoros, Congo, Costa Rica, Cote d'Ivoire, Croatia, Cuba, Cyprus, Democratic Republic of Congo, Djibouti, Dominican Republic, Ecuador, Egypt, El Salvador, Equatorial Guinea, Eritrea, Estonia, Eswatini, Ethiopia, Finland, Gabon, Gambia, Georgia, Ghana, Greece, Guatemala, Guinea, Guinea-Bissau, Guyana, Haiti, Honduras, Hungary, Iceland, Ireland, Jamaica, Japan, Jordan, Kazakhstan, Kenya, Kuwait, Kyrgyzstan, Latvia, Lebanon, Lesotho, Liberia, Libya, Lithuania, Luxembourg, Madagascar, Malawi, Malaysia, Maldives, Mali, Malta, Mauritania, Mauritius, Moldova," "Mongolia, Montenegro, Mozambique, Myanmar, Namibia, Nepal, New Zealand, Nicaragua, Niger, Nigeria, North, Acedonia, Norway, Oman, Palestine, Panama, Papua New Guinea, Paraguay, Qatar, Rwanda, Saint Lucia, Sao Tome and Principe, Saudi Arabia, Senegal, Serbia, Seychelles, Sierra Leone, Singapore, Slovakia, Slovenia, Somalia, South Korea, South Sudan, Sri Lanka, Sudan, Suriname, Syria, Taiwan, Tajikistan, Tanzania, Thailand, Togo, Trinidad and Tobago, Tunisia, Ukraine, United Arab Emirates, United States, Uruguay, Uzbekistan, Venezuela, Vietnam, Yemen, Zambia, Zimbabwe".

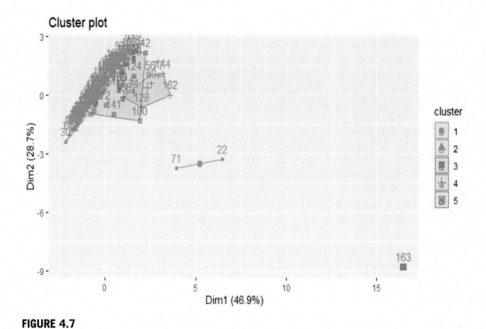

FIGURE 4.7

A cluster plot for the first and second principal components.

5. Cluster 5 is formed by 22 countries having the second least in total cases per million, reproduction rate, median age, total cases, new cases, life expectancy, and total deaths.

Fig. 4.7 shows a cluster plot for the first and second principal components. This plot reveals that cluster 2 is dense and some of the data points of cluster 2 are closer to cluster 5. Similarly, some of the data points of cluster 5 overlapped with cluster 4. There is no overlapping of clusters 1 and 3 with others. These two clusters with 2 and 1 data points are away from the other clusters. Thus, the study derives five clusters of sizes 2, 136, 1, 9, and 22 countries. Thus, K-means has the prospective to unveil the unsupervised machine learning technique for COVID-19 countries cluster analysis.

4.4 Time-series modeling for COVID-19 new cases

4.4.1 Time-series modeling: an overview

This section presents a prototype based on ML for the time-series analysis and forecasting of COVID-19 new cases in India. WHO's COVID-19 Situation Reports and data world, an open-source collaborative data community platform, was refereed for this study [33,40]. These reports provide worldwide details of the present COVID-19 pandemic situation and present anticipated case transmission classifications and death counts. The present research has simulated time-series prototyping of COVID-19 daily new infected patients in India. Time-series data are data points collected over some time as a sequence of time gaps. The daily news of COVID-19 cases in India during the period January 30, 2020, to December 30, 2020, was considered for the study. Time-series data analysis is carried out by analyzing this data to find out the pattern or trend of COVID-19 new cases to forecast future counts.

4.4.2 Dataset description

The dataset for the present study was retrieved from an open-access data platform named Our Complete COVID-19 [39]. This database represents a gathering of the COVID-19 data managed by Our World in Data. This dataset is on confirmed cases and deaths from Johns Hopkins University [35]. This dataset consists of worldwide confirmed cases, deaths, hospitalizations, and testing, as well as other variables of potential interest. The objective of the present study is time-series modeling, and new COVID-19 cases in India from January 30, 2020, to December 30, 2020, were considered. The dataset consists of COVID-19 new cases and daily updates from January 30, 2020, to December 30, 2020, with a total of 336 readings. These time-series data are plotted as shown in Fig. 4.8. The maximum count of new cases was found on September 16, 2020, that is, 97,894. It is observed from the plot that the count of new cases slowly enhanced in the

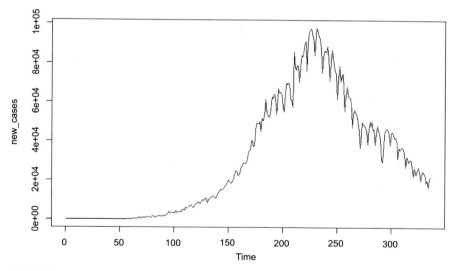

FIGURE 4.8

COVID-19 daily new cases in India.

opening phase, then raised exponentially, reached the peak, and then the count decreased. The statistical summary of this dataset is shown in Table 4.3.

4.4.3 Time-series exploration

The architecture for proposed time-series prototyping for forecasting COVID-19 new cases comprises four functions as shown below:

1. Time-series data visualization;
2. Plotting Auto Correlation Function (ACF) and Partial correlation function (PACF) graphs;
3. Developing time-series prototypes; and
4. Forecasting future COVID-19 infection spread cases.

A study of past trends has been analyzed for accurately predicting future trends and developing time-series models [41]. The available COVID-19 dataset was uploaded into the R working memory, and the data frame was transformed into the time-series class as needed for time-series prediction. The cycle time of this series was set to 1 because the database was having several daily COVID-19 fresh cases. The output of the augmented Dickey—Fuller test referred to in Fig. 4.9 revealed that this series was adequately stable to carry out different types of time-series prototyping. The correlation plots such as ACF and PACF plots are shown in Figs. 4.10 and 4.11, respectively. The blue line in the plot represents important unique values than "0". These plots helped in identifying the type of stationary series.

The time-series forecasting of new cases of COVID-19 was carried out by building theAutoregressive Integrated Moving Average (ARIMA) prototype "t series" package in R [42]. The discussion on the experiments conducted in the R platform is shown in this section. The forecasting of COVID-19 new cases was carried out using an autoregressive integrated moving average [43]. ARIMA prototype is fit by adding seasonal components, and the model is designed to forecast COVID-19 new cases. The forecasting results are shown in Fig. 4.12.

The green line indicates model fitness, and the blue region shows the prediction for the next few days. The prototype was summarized in Fig. 4.13. The

Table 4.3 Statistical summary of the dataset.

Facet	New cases
Minimum	0.0
First quartile	1556
Median	23,509
Mean	30,556
Third quartile	52,165
Maximum	97,894

```
Augmented Dickey-Fuller Test

data:  diff(x1)
Dickey-Fuller = -20.531, Lag order = 0, p-value = 0.01
alternative hypothesis: stationary
```

FIGURE 4.9

Augmented Dickey—Fuller test result.

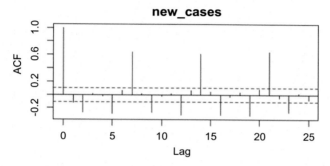

FIGURE 4.10

ACF plot of a daily update of new cases of COVID-19.

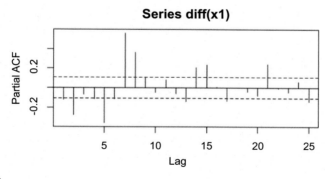

FIGURE 4.11

PACF plot of a daily update of new cases of COVID-19.

efficiency of this model was measured in Root Mean Squared Error (RMSE) and model fit.

Thus derived time-series model was used for the prediction of COVID-19 new cases in India for the period January 1 to January 18, 2021. As 0 new cases were

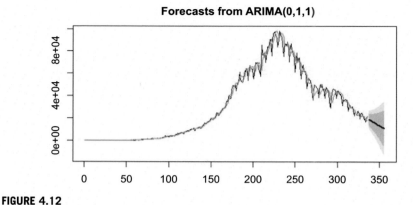

FIGURE 4.12

ARIMA prototype for new cases of COVID-19 in India.

```
Call:
arima(x = x1, order = c(0, 1, 1), seasonal = list(order = c(0, 1, 1), period = 1)

Coefficients:
        ma1    sma1
      -0.814  -0.814
s.e.   NaN     NaN

sigma^2 estimated as 15465838:   log likelihood = -3240.64,   aic = 6487.28

Training set error measures:
                   ME     RMSE     MAE MPE MAPE     MASE      ACF1
Training set -37.96708 3920.941 2203.78 NaN  Inf 0.9739958 0.2858482
```

FIGURE 4.13

ARIMA model summary.

found on January 8, it is skipped from forecasting. Table 4.4 compares the actual and predicted values of COVID-19 new cases. Fig. 4.14 compares the time-series model forecasted values with the actual values for test data. Both the RMSE and model fit reveal that the time-series modeling using ARIMA forecasts with higher accuracy.

4.4.4 Predictive analytics

Authors observed that COVID-19 new cases have gradually started increasing at the beginning of 2021. This time-series analysis was extended by training the model with additional data. The COVID-19 new cases update is collected as of March 20, 2021, and the time-series prototype is retrained with the additional information. Thus the dataset consists of COVID-19 new cases daily updates till March 20, 2021. The study was intended at predicting the number of new cases

Table 4.4 Forecasted and the real count of COVID-19 new cases in India.

Date	Actual count	Predicted count
January 1, 2021	20,035	18,515.84
January 02, 2021	37,256	18,074.63
January 03, 2021	16,504	17,633.42
January 04, 2021	16,375	17,192.21
January 05, 2021	18,088	16,750.99
January 06, 2021	20,346	16,309.78
January 07, 2021	18,139	15,868.57
January 09, 2021	36,867	15,427.35
January 10, 2021	16,311	14,986.14
January 11, 2021	12,584	14,544.93
January 12, 2021	15,968	14,103.72
January 13, 2021	16,946	13,662.5
January 14, 2021	15,590	13,221.29
January 15, 2021	15,158	12,780.08
January 16, 2021	15,144	12,338.87
January 17, 2021	13,788	11,897.65
January 18, 2021	10,050	11,456.44

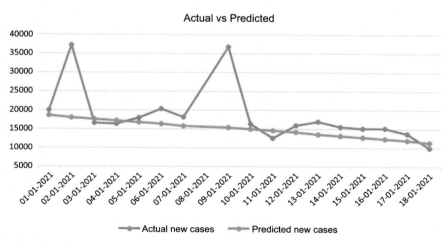

FIGURE 4.14

COVID-19 new cases actual count with model predicted count.

for April, May, and June. The date-wise forecasting of COVID-19 new cases for these three months is given in Table 4.5. This clearly shows the gradual increase in new cases of COVID-19 in India.

Table 4.5 Counts of new cases of COVID-19 predicted

April 2021		May 2021		June 2021	
April 01, 2021	58,963	May 01, 2021	106,126	June 01, 2021	154,861
April 02, 2021	60,535	May 02, 2021	107,698	June 02, 2021	156,433
April 03, 2021	62,107	May 03, 2021	109,270	June 03, 2021	158,006
April 04,2021	63,680	May 04, 2021	110,843	June 04, 2021	159,578
April 05, 2021	65,252	May 05, 2021	112,415	June 05, 2021	161,150
April 06, 2021	66,824	May 06, 2021	113,987	June 06, 2021	162,722
April 07, 2021	68,396	May 07, 2021	115,559	June 07, 2021	164,294
April 08, 2021	69,968	May 08, 2021	117,131	June 08, 2021	165,866
April 09, 2021	71,540	May 09, 2021	118,703	June 09, 2021	167,438
April 10, 2021	73,112	May 10, 2021	120,275	June 10, 2021	169,010
April 11, 2021	74,684	May 11, 2021	121,847	June 11, 2021	170,582
April 12, 2021	76,256	May 12, 2021	123,419	June 12, 2021	172,154
April 13, 2021	77,828	May 13, 2021	124,991	June 13, 2021	173,727
April 14, 2021	79,401	May 14, 2021	126,564	June 14, 2021	175,299
April 15, 2021	80,973	May 15, 2021	128,136	June 15, 2021	176,871
April 16, 2021	82,545	May 16, 2021	129,708	June 16, 2021	178,443
April 17, 2021	84,117	May 17, 2021	131,280	June 17, 2021	180,015
April 18, 2021	85,689	May 18, 2021	132,852	June 18, 2021	181,587
April 19, 2021	87,261	May 19, 2021	134,424	June 19, 2021	183,159
April 20, 2021	88,833	May 20, 2021	135,996	June 20, 2021	184,731
April 21, 2021	90,405	May 21, 2021	137,568	June 21, 2021	186,303
April 22, 2021	91,977	May 22, 2021	139,140	June 22, 2021	187,875
April 23, 2021	93,549	May 23, 2021	140,712	June 23, 2021	189,448
April 24, 2021	95,122	May 24, 2021	142,285	June 24, 2021	191,020
April 25, 2021	96,694	May 25, 2021	143,857	June 25, 2021	192,592
April 26, 2021	98,266	May 26, 2021	145,429	June 26, 2021	194,164
April 27, 2021	99,838	May 27, 2021	147,001	June 27, 2021	195,736
April 28, 2021	101,410	May 28, 2021	148,573	June 28, 2021	197,308
April 29, 2021	102,982	May 29, 2021	150,145	June 29, 2021	198,880
April 30, 2021	104,554	May 30, 2021	151,717	June 30, 2021	200,452
		May 31, 2021	153,289		

4.5 Conclusion

This chapter has two objectives: the first objective was to summarize the role of AI-based solutions in combating the COVID-19 pandemic. The second objective was to develop and experiment with an AI-based prototypes for predicting the trends of COVID-19 spread. The first part of the chapter presented an overview of the COVID-19 disease spread and its adverse impact on the global economy and public health. The review of various AI-powered solutions to fight against

COVID-19 and recent literature on AI-powered solutions from high-impact journals was presented. The authors have also presented AI-inspired two case studies as a part of COVID-19 analytics to fulfill the second objective of this chapter.

In the first case study, the identification of COVID-19 infected countries was carried out using an unsupervised learning approach. It was observed from the dataset that the highest number of new cases, that is, 862,181, was recorded on January 7, 2021. Hence, 170 countries' COVID-19 data as recorded on January 7, 2021, was considered for the present study. These clusters of countries are formed based on the parameters such as total cases per million, new cases, total cases and total deaths, median age, life expectancy, and reproduction rate. This study derived five clusters of sizes 2, 136, 1, 9, and 22 countries. Thus K-means clustering technique has the prospective to unveil the unsupervised learning technique for COVID-19 countries cluster analysis.

In the second case study, a machine learning prototype was designed for the time-series analysis and prediction of COVID-19 new cases in India. Time-series data and data points are collected over the same time as a sequence of time gaps. The count of daily new COVID-19 cases in India from January 30, 2020, to December 30, 2020, is within the scope of the study. Time-series data analysis was carried out by analyzing this data to find out the pattern or trend of COVID-19 new cases to forecast the future counts. It was observed from the analysis that the count of new cases slowly increased in the first phase, then raised exponentially, reached the peak, and then the count decreased. But later at the beginning of 2021, the COVID-19 new cases gradually started increasing again as the second wave of COVID-19 spread. The research is extended by training the model with COVID-19 new cases till March 20, 2021. This time-series model was used for predicting the number of new cases for April, May, and June 2021. This forecasting shows that there will be a gradual increase in COVID-19 new patients in India.

The research study demonstrates that the analysis of country clusters and forecasting of COVID-19 new cases can be handled by machine learning models. In this context, AI-based solutions such as analysis of COVID-19 country clusters and time-series forecasting are portrayed here. This research suggests that AI is being used as a tool to support the fight against the viral pandemic that has affected the entire world since the beginning of 2020 to help state administration to enact policies to control the pandemic situations from adversaries.

References

[1] K. Chriscaden, Communications Officer, World Health Organization, https://www.who.int/news/item/13-10-2020-impact-of-Covid-19-on-people's-livelihoods-their-health-and-our-foodsystems#:~:text = The%20COVID%2D19%20pandemic%20has,and%20the%20world%20of%20work.

[2] L. Jones, D. Palumbo, D. Brown, Coronavirus: How the pandemic has changed the world economy, https://www.bbc.com/news/business-51706225.

[3] P. Blake, D. Wadhwa, The impact of Covid-19 in 12 Charts, December 14, 2020, https://blogs.worldbank.org/voices/2020-year-review-impact-Covid-19-12-charts.

[4] J.S. Suri, A. Puvvula, et al., Covid-19 pathways for brain and heart injury in comorbidity patients: a role of medical imaging and artificial intelligence-based COVID severity classification: a review, Computers Biol. Med. 124 (2020) 103960. Available from: https://doi.org/10.1016/j.compbiomed.2020.103960.

[5] Z.I. Attia, S. Kapa, P.A. Noseworthy, et al., Artificial intelligence ECG to detect left ventricular dysfunction in Covid-19: a case series, Mayo Clin. Proc. 95 (11) (2020) 2464−2466. Available from: https://doi.org/10.1016/j.mayocp.2020.09.020.

[6] K. Zhang, X. Liu, J. Shen, et al., Clinically applicable AI system for accurate diagnosis, quantitative measurements, and prognosis of Covid-19 pneumonia using computed tomography, Cell 181 (2020) 1423−1433. Available from: https://doi.org/10.1016/j.cell.2020.08.029. e1−e11.

[7] H. Greenspan, R.S. JoséEstépar, et al., Position paper on Covid-19 imaging and AI: from the clinical needs and technological challenges to initial AI solutions at the lab and national level towards a new era for AI in healthcare, Med. Image Anal. 66 (2020) 101800. Available from: https://doi.org/10.1016/j.media.2020.101800.

[8] O.S. Albahria, A.A. Zaidana, A.S. Albahrib, et al., Systematic review of artificial intelligence techniques in the detection and classification of Covid-19 medical images in terms of evaluation and benchmarking: taxonomy analysis, challenges, future solutions and methodological aspects, J. Infect. Public Health 13 (2020) 1381−1396. Available from: https://doi.org/10.1016/j.jiph.2020.06.028.

[9] S. Lalmuanawma, J. Hussain, L. Chhakchhuak, Applications of machine learning and artificial intelligence for Covid-19 (SARS-CoV-2) pandemic: a review, Chaos Solitons Fract. 139 (2020) 110059. Available from: https://doi.org/10.1016/j.chaos.2020.110059.

[10] D.N. Vinod, S.R.S. Prabaharan, Data science and the role of artificial intelligence in achieving the fast diagnosis of Covid-19, Chaos Solitons Fract. 140 (2020) 110182. Available from: https://doi.org/10.1016/j.chaos.2020.110182.

[11] J. Rasheed, A. Jamil, A.A. Hameed, et al., A survey on artificial intelligence approaches in supporting frontline workers and decision-makers for the Covid-19 pandemic, Chaos Solitons Fract. 141 (2020) 110337. Available from: https://doi.org/10.1016/j.chaos.2020.110337.

[12] M.-H. Tayarani N., Applications of artificial intelligence in battling against Covid-19: a literature review, Chaos Solitons Fract. 142 (2021) 110338. Available from: https://doi.org/10.1016/j.chaos.2020.110338.

[13] A. Imran, I. Posokhov, H.N. Qureshi, et al., AI4Covid-19: AI-enabled preliminary diagnosis for Covid-19 from cough samples via an app, Inform. Med. Unlocked 20 (2020) 100378. Available from: https://doi.org/10.1016/j.imu.2020.100378.

[14] B. Wanga, S. Jin, Q. Yan, H. Xu, et al., AI-assisted CT imaging analysis for Covid-19 screening: building and deploying a medical AI system, Appl. Soft Comput. J. 98 (2021) 106897. Available from: https://doi.org/10.1016/j.asoc.2020.106897.

[15] S.R. Nayak, D.R. Nayak, et al., Application of deep learning techniques for detection of Covid-19 cases using chest X-ray images: a comprehensive study, Biomed. Signal. Process. Control. 64 (2021) 102365. Available from: https://doi.org/10.1016/j.bspc.2020.102365.

[16] A.C. Kaushik, U. Raj, AI-driven drug discovery: a boon against Covid-19, AI Open. 1 (2020) 1−4. Available from: https://doi.org/10.1016/j.aiopen.2020.07.001.

[17] F. Dorr, H. Chaves, et al., Covid-19 pneumonia accurately detected on chest radiographs with artificial intelligence, Intelligence-Based Med. 3–4 (2020) 100014. Available from: https://doi.org/10.1016/j.ibmed.2020.100014.

[18] A.C. Chang, Artificial intelligence and Covid-19: present state and future vision, Intelligence-Based Med. 3–4 (2020) 100012. Available from: https://doi.org/10.1016/j.ibmed.2020.100012.

[19] Y. Zhou, F. Wang, J. Tang, R. Nussinov, F. Cheng, Artificial intelligence in Covid-19 drug repurposing, http://www.thelancet.com/digital-health. Vol. 2 December 2020. Available from: https://doi.org/10.1016/S2589-7500(20)30192-8.

[20] R. Vaishya, M. Javaid, I.H. Khan, A. Haleem, Artificial intelligence (AI) applications for Covid-19 pandemic, Diabetes Metab. Syndr. 14 (2020) 337–339. Available from: https://doi.org/10.1016/j.dsx.2020.04.012.

[21] W. Naudé, Artificial intelligence vs COVID 19: limitations, constraints and pitfalls. Available from: https://doi.org/10.1007/s00146-020-00978-0.

[22] A.J. DeGrave, J.D. Janizek, and S.-I. Lee, AI for radiographic Covid-19 detection select shortcuts over signal. Available from: https://doi.org/10.1101/2020.09.13.20193565.

[23] Z. Yang, Z. Zeng1, K. Wang, S.-S. Wong, et al., Modified SEIR and AI prediction of the trend of the epidemic of Covid-19 in China under public health interventions.

[24] A.A. Hussain, O. Bouachir, F. Al-Turjman, AI Techniques for Covid-19, Digital Object Identifier https://doi.org/10.1109/ACCESS.2020.3007939, IEEE access Artificial Intelligence (AI) and Big Data for Coronavirus (Covid-19) Pandemic.

[25] Q. Pham, D. Nguyen, T. Huynh-The, W.-J. Hwang, and P.N. Pathirana, A Survey on the State-of-the-Arts. Available from: https://doi.org/10.1109/ACCESS.2020.DOI, IEEE Access.

[26] Z. Allam, G. Dey, D.S. Jones, Artificial intelligence (AI) provided early detection of the coronavirus (Covid-19) in China and will influence future urban health policy internationally, AI 1 (2020) 156–165. Available from: https://doi.org/10.3390/ai1020009.

[27] S. Girard, Using artificial intelligence to help combat Covid-19, Follow the latest Covid-19 developments in your country with real-time AI-powered news and data at http://www.oecd.ai/covid.

[28] T.T. Nguyen, AI in the battle against coronavirus. https://doi.org/10.13140/RG.2.2.36491.23846/1, arXiv:2008.07343v1 [cs.CY] 30 Jul 2020.

[29] L. Wynants, Prediction models for diagnosis and prognosis of covid-19: a systematic review and critical appraisal, Published online 2020 Apr 7. https://doi.org/10.1136/BMJ.m1328, BMJ. 2020; 369: m1328, BMJ Publishing Group, https://www.ncbi.nlm.nih.gov/pmc/articles/PMC7222643/.

[30] OECD Website (Red for Latest Updates on Corona), https://www.oecd.org/coronavirus/en/.

[31] S. Dananjayan, G.M. Raj, Artificial Intelligence during a pandemic: The COVID-19 example. Int. J. Health Plann. Manage. 35 (5) (2020) 1260–1262. Available from: https://doi.org/10.1002/hpm.2987. Epub 2020 May 20. PMID: 32430976; PMCID: PMC7276785.

[32] B. Asmika, A study on Artificial Intelligence in the Times of Covid-19, ISSN: 2454-9150 Vol-06, Issue-03, June 2020 227 I IJREAMV06I0363092 https://doi.org/10.35291/2454-9150.2020.0469, International Journal for Research in Engineering Application & Management (IJREAM) ISSN: 2454-9150, vol-06, Issue-03, June 2020.

[33] B. McCall, Covid-19 and artificial intelligence: protecting health-care workers and curbing the spread. Available from: https://doi.org/10.1016/S2589-7500(20)30054-6, https://doi.org/10.1128/JVI.00127-20.

[34] Coronavirus Disease (Covid-19) Situation Reports, https://www.who.int/emergencies/diseases/novel-coronavirus-2019/situation-reports, retrieved on 17th Jan 2021.

[35] J. Hasell, E. Mathieu, D. Beltekian, B. Macdonald, C. Giattino, E. Ortiz-Ospina, et al., A cross-country database of Covid-19 testing, Sci. Data 7 (2020). Available from: https://doi.org/10.1038/s41597-020-00688-8. retrieved on 17th Jan 2021.

[36] H. Ritchie, Coronavirus Source Data https://ourworldindata.org/coronavirus-source-data, retrieved on 18th Jan 2021.

[37] Data on Covid-19 (coronavirus) by Our World in Data, https://github.com/owid/Covid-19-data/tree/master/public/data, retrieved on 18th Jan 2021.

[38] R.S. Kamath, S.S. Jamsandekar, K.G. Kharade, R.K. Kamat, Data Analytics in R: A Case Study Based Approach, Himalaya Publishing House Pvt. Ltd, 2019. ISBN: 978-93-5367-791-6.

[39] R.S. Kamath, R.K. Kamat, Visualization of University clusters based on NIRF and NAAC scores: K-means algorithm approach, University News, A Wkly. J. High. Educ. 57 (03) (2019). ISSN: 05662257.

[40] R.K. Kamat, R.S. Kamath, Visualization of earthquake clusters over space: K-means approach, J. Chem. Pharm. Sci. (JCHPS) 10 (1) (2017) 250–253. ISSN: 0974-2115.

[41] S. Pati, Covid_19_India dataset, https://data.world/gabbarsingh/covid19india retrieved on 3rd Feb 2021.

[42] Srivastava, T., A Complete Tutorial on Time Series Modelling in R, https://www.analyticsvidhya.com/blog/2015/12/complete-tutorial-time-series-modeling/ retrieved 25th August 2018.

[43] Dalinina, R., Introduction to Forecasting with ARIMA in R, https://www.datascience.com/blog/introduction-to-forecasting-with-arima-in-r-learn-data-science-tutorials August 27 2018.

Further reading

R.S. Kamath, R.K. Kamat, Time-series analysis and forecasting of rainfall at Idukki district, Kerala: machine learning approach, Disaster Adv. 11 (11) (2018) 27–33.

Graph convolutional networks for pain detection via telehealth

5

Suzan Anwar[1,2], Mariofanna Milanova[3], Shereen Adbulla[4] and Saja Ataallah Muhammed[5]

[1]Computer Science Department, Philander Smith College, Little Rock, AR, United States
[2]Computer Science Department, Salahaddin University, Erbil, Iraq
[3]Computer Science Department, University of Arkansas, Little Rock, AR, United States
[4]Computer Science Department, Ploy Tech University, Erbil, Iraq
[5]Computer Science and Information Technology Department, Salahaddin University, Erbil, Iraq

5.1 Introduction

In 2020, a severe crisis was created because of the highly contiguous COVID-19 and its fast spread [1]. The public's concerns are increased due to the absence of effective treatment for this pandemic [2]. Around the world, the healthcare delivery systems were significantly impacted by COVID-19. In early 2020, the Secretary of Health and Human Services declared a public health emergency and allowed beneficiaries to receive telehealth services in their home [3]. The need to detect pain during telehealth is increased to help the physician improve outcomes for outpatients and more control back to them with more appropriate medications [4].

Deep learning is a crucial machine learning tool to solve complex problems and achieve supervised learning. Many ways are using different measures to describe pain levels such as speech, physiological, facial expression, and body gesture.

The long short-term memory (LSTM) and neural networks (NNs) to detect pain from speech analysis are used in Ref. [5] and consider the first research to use speech for pain detection. The vocal features are expected using Chinese corpus, and the unsupervised learning NN is employed in the first layer. To obtain the sentence-level acoustic representation as an output for each patient, the vocal features are fine-tuned using a triage dataset and NNs. Finally, a support vector machine (SVM) is used in the classes to detect pain levels. The experimental findings showed that the proposed method achieved 72.3% weight average recall in two classes.

A physiological signal is one of the viral approaches for pain assessments. It depends on the physiological response of the patient's body. Brain dynamic or vital signals such as muscle activity, heart rate, and blood pressure are examples of physiological signals. A model using NN techniques is used in Ref. [6]. The author in Ref. [7] proposed a multimodel data system that used signals from

Artificial Intelligence in Healthcare and COVID-19. DOI: https://doi.org/10.1016/B978-0-323-90531-2.00006-0

electrocardiograms (ECGs) and skin conductance. To build and conduct their model, the BioVid Heat Pain dataset is used. Then, they compared it with other available machine learning techniques such as SVM with radial basis function kernel, SVM with linear kernel, and logistic regression. The resulting performance was around 82.75%.

Recently, signals from facial expressions are employed by researchers for many applications such as face recognition. In Ref. [8], the pain is detected from facial expressions using deep learning. In the first stage, the authors used recurrent neural network (RNNs) to calculate Prkachin and Solomon's pain intensity (PSPI) for each video frame. In the second stage, the authors utilized hidden conditional random fields that used the estimated output from the first stage to predict the patient's self-reported visual analog scale. The proposed approach achieved a high performance compared with other available methods.

Another study done by Ref. [9] used deep learning to detect binary pain (pain, no pain) based on facial expression. In the first stage, the authors used convolutional neural networks (CNNs) for feature extraction from faces. In the second stage, they train the LSTM using the features map from the previous stage. The UNBC-McMaster dataset is utilized for their experiments, and the results showed high performance achieving 93.3%.

In 2017, a pain estimation model was proposed by Ref. [10]. The model combined handcrafted features obtained from face images with features extracted from deep learning using CNNs. They classify the level of pain using the linear regression model. The proposed method's experimental results indicated the limitation in its performance where the root mean square error was 0.99.

The research in Ref. [11] aimed to use the WebFace dataset, and deep learning for pain estimation is proposed. The authors applied regression loss with the center loss. To evaluate their method, new metrics are presented. The performance of the proposed method scored a high performance using mean absolute error with 0.389.

In 2018, a study in Ref. [12] utilized trained CNN to obtain cumulative attributes (CA) as the first stage. In the second stage, the output is produced using the trained regression model. The proposed pain estimation model results achieved high performance when compared to non-CA-CNN. A new RGB, depth, and thermal images dataset is built to recognize the pain level in recorded videos. The constructed dataset included 20 healthy people who induced pain with electrical pulses. The participant subjects were asked to determine the five levels of pain during the pain simulation process. After building the video dataset, the authors constructed a pain detection model using deep learning and spatiotemporal features.

The new method in Ref. [13] used 2D CNN to extract the image's facial features to recognize pain. Then, the author found the relation between sequence pain levels and frames using LSTM. The fusion matrix showed the most outstanding performance of the proposed method compared with other modalities.

In 2017, a method in Ref. [14] to recognize lower back pain was suggested. The authors use a deep learning approach to extract the features from tomography lumbar spine X-ray images with five vertebral levels (normal, mild, crush, wedge, and severe). Recently, a method to detect low back pain using a deep learning method is proposed [15]. Three motion sensors were attached to human skin to obtain kinematic data as input to the LSTM network. The experimental findings showed a high accuracy of 97.2%.

The body movement dataset called em-pain is used by Ref. [16]. The authors applied LSTM to the attention mechanism to assess the pain from the behavior. A sliding window with zero paddings is utilized for data segmentation, after comparing the proposed method to the previous techniques obtaining a high performance of 0.844 mean F1 scores.

Several CNN frameworks applied on the BioVid heat pain dataset are explored in Ref. [17]. ECG, EDA, and EMG modalities of signals are used as input of 1D and 2D formats. The findings of the proposed method reached 84.57% for binary (two classes) classification. In 2019, a study to recognize the chronic pain behavior of LBP patients using deep learning was proposed in Ref. [18]. The authors applied two RNNs, dual LSTM and stacked LSTM. They were able to calculate the energies and angles of mockup data from five muscle activities only. After applying different experiments for each muscle activity, the performance reached 75%.

In Ref. [19], the pain is detected using the facial action coding system (FACS), which defines facial muscle action units (AUs). The authors processed the facial videos frame by frame to calculate the AUs likelihood values using a deep NN. Their results showed improved performance significantly when compared with other pain detection methods.

In [20], the authors built a deep NN architecture for competitive models to recognize pain levels using the BioVid dataset. The work aims to assess physiological channels and generate an accurate classification. The average performance achieved 84.57% for the binary classification.

In this chapter, we propose a new pain detection model for video-based datasets, and our contributions are summarized as follows:

- This chapter's proposed model is the first research that applies graph convolutional networks (GCNs) to recognize pain levels during telehealth.
- In this chapter, fully automated effective machine learning and computer vision pain detection are developed.
- The developed model stores the result in a CSV file for research and clinical purposes.
- We evaluated the proposed model by utilizing two datasets, and we added several data columns as an extra input to obtain better results.
- The features of each video's frames are weighted using the GCN layer's adjacency matrix to generate different pain intensities of each frame.

The rest of the chapters' sections are organized as follows: Section 5.2 describes our model, including extracting pain features, GCN layers, and classification. Section 5.3 depicts the analysis of the experimental. Section 5.4 discusses the results of evaluating the method's performance. Section 5.5 concludes the chapter.

5.2 Methodology

The proposed model consists of four layers. The first layer is features extraction using the CNN layer that extracts the facial pain features from a given video sequence x_i, $i = 1, 2, 3, \ldots, N$, where N is the total number of frames in each video. The second layer is the graph CNN layer that learns more attractive facial pain features of each frame by applying GCN and LSTM. The third layer is the weighted feature fusion layer to add weights to each feature yield from the previous layer. The fourth layer is the classification layer that receives the N features and detects the pain. Fig. 5.1 shows the overall architecture pipeline of the proposed model, which is explained in the following sections.

5.2.1 Features extraction

After generating the N frame features using CNN for the first layer, N nodes for each frame are built in the learnable adjacency matrix of the GCN layer. All neighbor nodes share and update their features dynamically with backpropagation gradient and from the last layer matrix. Simultaneously, the nodes update and propagate their state and messages with neighbor nodes in each layer to generate interdependency among the frames. The weak pain frames usually depend on the peak frames and focus on the next pain region. Each node updates its message from the nearest peak frame and concentrates on the interesting pain region. The neighbors' features are embedded with the adjacency matrix, while messages are updated and then propagated to nodes. Each node updates its state from its current state and updates messages in the adjacency correlation matrix.

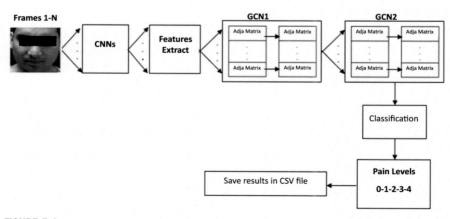

FIGURE 5.1

The overall architecture pipeline.

A matrix $m_i \in R^{(N-1) \times d}$ that holds the features H can be represented jointly as follows, where i is the ith node that is connected to $N-1$ other neighbors nodes to receive the messages, and d is the node's dimension.

$$m_i = \left[H_1^T H_2^T H_3^T \ldots H_{i-1}^T H_{i+1}^T \ldots H_N^T\right]^T \tag{5.1}$$

The embedded messages of the neighbor's nodes E_i^l are computed from the embedded features with a learnable parameter matrix $W^l \in R^{d \times d}$ while updating the messages and propagating the node i.

$$E_i^l = m_i W^l \tag{5.2}$$

The output of the current ith node $o_i^{l+1} \in R^{1 \times d}$ can be computed by adding the correlation coefficients matrix A_{ii} between the current node and its neighbors, embedded messages from Eq. (5.1), and the current state of the node itself in the dynamic adjacency matrix using nonlinear function LeakyRelLu.

$$o_i^{l+1} = f(A_{\underline{ii}} M_i^l \oplus A_{ii} H_i W^l) \tag{5.3}$$

To update the trained adjacency matrix A, the loss is obtained with the learning rate lr, and the backpropagation is conducted with gradients in the graph.

$$A^{l+1} = A^l - lr * \partial loss / \partial A^l \tag{5.4}$$

5.2.2 Graph-based modules

The frame's updated features resulting from the previous stage mainly focus on the most contributing facial pain areas. The long-term BiLSTM temporal is used to access past and future states information for combining and computing each frame's feature during the time step to produce the output learned feature of each frame using the following equation:

$$H_i^{l+1} = g(V_{f\sigma}(U_f[s_f^l, o_i^{l+1}]) + V_{b\sigma}(U_b[s_b^l, o_i^{l+1}]) + b), i \in [1, N] \tag{5.5}$$

The information from the past and the next time steps of the hidden states is represented as s_f^l and s_b^l, respectively. Both input and hidden states are embedded in two directions and are represented as U_f and U_b, respectively. The activation function tan$h$ is represented as g, the bias as b, and the sigmoid function as σ.

The adjacency matrix A is initiated with elements containing 1 for its main diagonal and 0 for the rest. Each frame dependency is updated via a learnable GCN layer. The out of the GCN layer is used to explore the long-term dependency in N steps during the LSTM layer. The deep features are constructed using two GCN modules and shared adjacency matrix A. The graph model focuses on learning the pain features in the most contributing pain regions known as peak frame, while some features exist in the weak frame. The GCN layer guides the video frames to find the most contributing pain region in neighbor frames by sharing features among them. Fig. 5.2 shows a cell in Jupyter notebook with Python graph class code to create both GCN layers.

5.2.3 Frame-wise weight calculation

In the previous layer, two GCN modules are used to obtain the learned features that have information to focus on the facial area for pain. The weak pain features in some frames do not have enough information because their intensities are less than the coefficients of the peak frame in the same video. Therefore a weight function for the adjacency matrix A is developed to compute each frame's weight. A row-wise average pooling and a softmax function are applied to the adjacency matrix A to obtain the pain intensity in each frame as a normalized weight.

$$\text{weight} = \text{softmax}(\text{mean}(A, dim = 0)) \tag{5.6}$$

The gradient of adjacency matrix A is frozen to achieve correct graph learning in the above weight function. The value of matrix A represents pain intensities. The pain intensity weights represent each video frame's pain intensity, where the peak frames assign larger weights than the weak frames. The weights are normalized for better understanding and visualization using the sigmoid function.

5.2.4 Classification

The contribution of the peak frames that have more informative features is reemphasized during the classification stage. The features H in each video frame are fused with the pain intensity weight of other frames to produce the final representation r by computing the sum of the features H and the weight.

$$r = \sum_{i=0}^{N} \text{weight}_i H_i \tag{5.7}$$

The frames with weak and peak pain features are distinguished with the previous layer's help via the weight fusion function. This increased the peak pain

```
In [ ]:  class GCN(nn.Module):
             def __init__(self, in_feats, hidden_size, num_classes):
                 super(GCN, self).__init__()
                 self.gcn1 = GCNLayer(in_feats, hidden_size)
                 self.gcn2 = GCNLayer(hidden_size, num_classes)
                 self.softmax = nn.Softmax()

             def forward(self, g, inputs):
                 h = self.gcn1(g, inputs)
                 h = torch.relu(h)
                 h = self.gcn2(g, h)
                 h = self.softmax(h)
                 return h
```

FIGURE 5.2

Python code for GCN layers. *GCN*, graph convolutional network.

features' contribution in the final classification step while decreasing the impact's weak pain features.

5.3 Experiments

Two widely used pain datasets are utilized to conduct the experiments. These datasets are the BioVid Heat Pain dataset [21,22] and UNBC-McMaster [23]. A detailed comparison between the used experimental datasets is shown in Table 5.1. A 10-fold cross-validation is employed on both datasets after performing the standard evaluation procedures. The exact split of test/train is used for each stage. To test and evaluate the proposed model's performance in this chapter, comparisons with the state-of-the-art approaches are made in the following sections.

5.3.1 Datasets

The following datasets are used to evaluate the proposed model:

- BioVid Heat Pain dataset: The videos of 90 subjects were collected as a study in Refs. [21,22]. A thermal stimulator was used to induce pain in the participants' arm (right). Cameras, videos, and physiological sensors are used to record the participants' facial pain responses. The response is saved as precise data within the value and time domain. The subject was selected with stimulation heat that produces the same pain intensity across other subjects. Four pain intensities are defined: PL1 as lowest heat that the subject defined as a pain; Pl4 as the highest pain intensity that the subject can experience; PL2 and PL3 as intermediate intensities defined in between. For 20 times, different heat stimulation was applied for 4 seconds and paused for 8−12 seconds. BioVid dataset includes five parts (A, B, C, D, and E), each part consisting of thousands of samples of tens of subjects, each with around 5 seconds of videos.
- UNBC-McMaster Shoulder Pain dataset

Table 5.1 Comparison between experimental datasets.

Attribute	UNBC-McMaster	BioVid
No. of subjects	129	90
Subject's type	Self-identified pain patient	Healthy volunteers
Pain type	Natural shoulder pain	Stimulated heat pain
Pain levels	0−16 (PSPI) and 0−10 (VAS)	1−4 (stimuli)
Size of the dataset	200 videos with 31,571 frames	17,300 5s videos with 25 fps

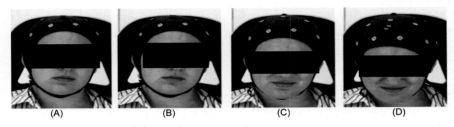

FIGURE 5.3

A sample of BioVid dataset with different pain levels, (A) pain level 1, (B) pain level 2, (C) pain level 3, (D) pain level 4.

The UNBC-McMaster Shoulder Pain dataset consists of 200 video sequences, and each video includes facial emotions for patients suffering from shoulder pain. It has around 48,398 farmers who are coded using the FACS coding system. Each video frame was rated by more than two coding experts, and it has an associated pain self-report. Each frame's detected face had been tracked with 66 points using active appearance model (AAM) [24]. The UNBC-McMaster dataset consists of 129 subjects, 66 of them are female, and 63 are male. All subjects are identified as suffering from shoulder problems and having genuine pain. Three physiotherapy clinics collected the information from the participant and performed motion tests: flexion, internal and external rotation, and abduction [25,26]. Fig. 5.3 shows a subject from the BioVid dataset experiencing four different pain levels.

5.3.2 Experimental setting

In our model, we choose N frames from each video. If the numbers of frames are less than N in the video, we reuse frames to reach N. Pain features are extracted using a pretrained model. The vector dimension d of the features is set to 256, and the nonlinear activation function for the GCN layer with a negative slope of 0.2 is adopted along with the BILSTM layer. Each video frame is cropped to detect the face and resized to 256 × 265 with frame flipping for data augmentation and changing the illumination. All experimental and training stages are conducted using PyTorch via Watson Studio and Quad NVIDIA Tesla 32GB V100 provided by IBM.

5.4 Results and discussion

The datasets BioVid and UNBC-McMaster are used for evaluation. We compare our model with state-of-the-art methods, which use video-based datasets. The results of testing the new model on the UNBC-McMaster dataset are compared with other state-of-the-art methods in Refs. [27−30] and [23]. As shown in

Table 5.2 Overall performance of accuracy on the experimental UNBC dataset.

Method	Accuracy
TSAFEW [27]	98.4%
External MTL [28]	95%
DeepFaceLIFT [29]	97.82%
pRNN-HCRF [30]	97.5%
Lucey et al. [23]	80.99%
Proposed	99.08%

Table 5.3 Confusion matrix for four pain level detections on UNBC dataset.

	PA0	PA1	PA2	PA3	PA4
PA0	100%	0%	0%	0%	0%
PA1	0%	100%	0%	0%	0%
PA2	0%	0%	100%	0%	0%
PA3	0%	0%	0%	99%	0%
PA4	0%	0%	0%	0%	100%

Table 5.4 Overall performance of accuracy on the experimental BioVid dataset.

Method	Accuracy
MT-NN [6]	82.57%
LBP-BSIF [31]	64.80%
RTP DFE [32]	56.75%
Head movement [21]	67.00%
Time windows [21]	71.00%
Proposed	89.3%

Table 5.2, our model achieved the best performance and achieved a 99.08% accuracy rate. It outperformed the methods in Refs. [27−30] and [23] by 0.68%, 4.08%, 1.26%, 1.58%, and 18.09%, respectively. These methods are video-based pain detection methods in our model. Table 5.3 represents the confusion matrix and shows that all pain levels are recognized well, except pain level 3 giving a recognition rate of 99%.

Table 5.4 reports the comparison of our model with other state-of-the-art methods represented in Refs. [6,31,32] and [21] on the BioVid dataset. Our model's pain recognition rates reached 89.3% and outperformed the four compared video-based pain detection methods and showed an improvement by 6.73%, 24.5%, 32.55%, and 18.3%, respectively. The confusion matrix of pain level

Table 5.5 Confusion matrix for four pain level detections on the BioVid dataset.

	PA0	PA1	PA2	PA3	PA4
PA0	100%	0%	0%	0%	0%
PA1	0%	100%	0%	0%	0%
PA2	0%	0%	77%	0%	0%
PA3	0%	0%	0%	68%	0%
PA4	0%	0%	0%	0%	100%

recognition on the BioVid dataset is represented in Table 5.5, which shows that the proposed model recognized pain levels 0, 1, and 4 accurately, while pain levels 2 and 3 had low recognition rates.

5.5 Conclusion

This chapter presents a novel approach to pain recognition via telehealth systems during the COVID-19 pandemic. We used graphs for learning about pain feature recognition. The features of each node are learned using graphs that are designed to propagate features from peak frames. An adjacency matrix is applied after learning from the graph layers to locate each video sequence's peak frame. Two pain video-based datasets are utilized to test the performance of the proposed model. The experimental findings of comparing our model with other state-of-the-art methods demonstrate our model's superiority in recognizing all pain levels, primarily with a 99.08 recognition rate.

Acknowledgment

We thank the authors in Refs. [21] and [24] for collecting BioVid and UNBC-McMaster datasets, National Science Foundation under Award No. OIA-1946391, DART Research Program.

References

[1] Y.R. Guo, Q.D. Cao, Z.S. Hong, Y.Y. Tan, S.D. Chen, H.J. Jin, et al., The origin transmission and clinical therapies on coronavirus disease 2019 (COVID-19) outbreak—an update on the status, Military Med. Res. 7 (1) (2020) 1—10.
[2] M. Cascella, M. Rajnik, A. Cuomo, S.C. Dulebohn and R. Di Napoli, Features Evaluation and Treatment Coronavirus (COVID-19), Stat Pearls, March 2020.
[3] H.A. Rothan, S.N. Byrareddy, The epidemiology and pathogenesis of coronavirus disease (COVID-19) outbreak, J. Autoimmunity 109 (2020) 102433.

[4] July 2020, [online]. Available from: https://edit.cms.gov/files/document/medicare-tel-ehealth-frequentlyasked-questions-faqs-31720.pdf.

[5] F.-S. Tsai, Y.-M. Weng, C.-J. Ng, C.-C. Lee. Embedding stacked bottleneck vocal features in a LSTM architecture for automatic pain level classification during emergency triage. Affective Computing and Intelligent Interaction (ACII), San Antonio, TX, USA, 23–26 October 2017; pp. 313–318.

[6] D. Lopez-Martinez, R. Picard. Multi-task neural networks for personalized pain recognition from physiological signals. Affective Computing and Intelligent Interaction Workshops and Demos (ACIIW), San Antonio, TX, USA, 23–26 October 2017; pp. 181–184.

[7] D.L. Martinez, O. Rudovic, R. Picard. Personalized Automatic Estimation of Self-Reported Pain Intensity from Facial Expressions. arXiv 2017, arXiv:1706.07154. Available online: http://arxiv.org/abs/1706.07154 (accessed on 26 July 2018).

[8] P. Rodriguez, G. Cucurull, J. Gonzalez, J.M. Gonfaus, K. Nasrollahi, T.B. Moeslund, et al., Deep pain: exploiting long short-term memory networks for facial expression classification, IEEE Trans. Cybern (2017) 1–11.

[9] J. Egede, M. Valstar, B. Martinez. Fusing Deep Learned and Handcrafted Features of Appearance, Shape, and Dynamics for Automatic Pain Estimation. 2017 12th IEEE International Conference on Automatic Face & Gesture Recognition (FG 2017), Washington, DC, USA, 30 May–3 June 2017; pp. 689–696.

[10] F. Wang, X. Xiang, C. Liu, T.D. Tran, A. Reiter, G.D. Hager, et al. Regularizing face verification nets for pain intensity regression. In Proceedings of the 2017 IEEE International Conference on Image Processing (ICIP), Beijing, China, 17–20 September 2017; pp. 1087–1091.

[11] S. Jaiswal, J. Egede, M. Valstar. Deep Learned Cumulative Attribute Regression. 2018 13th IEEE International Conference on Automatic Face & Gesture Recognition (FG 2018), Xi'an, China, 15–19 May 2018; pp. 715–722.

[12] M.A. Haque, R.B. Bautista, F. Noroozi, K. Kulkarni, C.B. Laursen, R. Irani, et al. Deep Multimodal Pain Recognition: A Database and Comparison of Spatio-Temporal Visual Modalities. In Proceedings of the 2018 13th IEEE International Conference on Automatic Face & Gesture Recognition (FG 2018), Xi'an, China, 15–19 May 2018; pp. 250–257.

[13] M. Bellantonio, M.A. Haque, P. Rodriguez, K. Nasrollahi, T. Telve, S. Escalera, et al., Spatio-temporal pain recognition in CNN-based super-resolved facial images, Image Processing, Computer Vision, Pattern Recognition, and Graphics, 10165, Springer, Cham, Switzerland, 2016, pp. 151–162.

[14] K.R. Kulkarni, A. Gaonkar, V. Vijayarajan, K. Manikandan, Analysis of lower back pain disorder using deep learning, IOP Conf. Series Mater. Sci. Eng., 263, 2017, p. 42086.

[15] B. Hu, C. Kim, X. Ning, X. Xu, Using a deep learning network to recognize low back pain in static standing, Ergonomics 61 (2018) 1374–1381.

[16] C. Wang, M. Peng, T.A. Olugbade, N.D. Lane, A.C.D.C. Williams, N. Bianchi-Berthouze. Learning Bodily and Temporal Attention in Protective Movement Behavior Detection. arXiv 2019, arXiv:1904.10824. Available from: http://arxiv.org/abs/1904.10824 (accessed on 5 June 2020).

[17] P. Thiam, P. Bellmann, H. Kestler, F. Schwenker, Exploring deep physiological models for nociceptive pain recognition, Sensors 19 (2019) 4503.

[18] C. Wang, T.A. Olugbade, A. Mathur, A.C.D.C. Williams, N.D. Lane, N. Bianchi-Berthouze. Recurrent network-based automatic detection of chronic pain protective behavior using MoCap and sEMG data. 23rd International Symposium on Wearable Computers, London, UK, 9−13 September 2019; pp. 225−230.

[19] G. Menchetti, Z. Chen, D.J. Wilkie, R. Ansari, Y. Yardimci, A.E. Çetin, Pain Detection from Facial Videos Using Two-Stage Deep Learning, 2019 IEEE Global Conference on Signal and Information Processing (GlobalSIP), Ottawa, ON, Canada, 2019, pp. 1−5. Available from: https://doi.org/10.1109/GlobalSIP45357.2019.8969274.

[20] P. Thiam, P. Bellmann, H. Kestler, F. Schwenker, Exploring deep physiological models for nociceptive pain recognition. Sensors 19 (20) (2019) 1424−8220.

[21] S. Walter, S. Gruss, H. Ehleiter, J. Tan, H.C. Traue, S. Crawcour, et al. The biovid heat pain database data for the advancement and systematic validation of an automated pain recognition system. In Proceedings of the 2013 IEEE International Conference on Cybernetics (CYBCO), Lausanne, Switzerland, 13−15 June 2013; pp. 128−131.

[22] P. Werner, A. Al-Hamadi, R. Niese, S. Walter, S. Gruss, & H.C. Traue, Towards Pain Monitoring: Facial Expression, Head Pose, a new Database, an Automatic System and Remaining Challenges. In Proceedings of the British Machine Vision Conference (BMVC). BMVA Press, 2013, pp. 119.1−119.13.

[23] P. Lucey, J.F. Cohn, K.M. Prkachin, P.E. Solomon, I. Matthews. Painful data: The UNBC-McMaster shoulder pain expression archive database. In Face and Gesture; IEEE: Santa Barbara, CA, USA, 2011; pp. 57−64.

[24] T. Cootes, G. Edwards, C. Taylor, Active appearance models, IEEE Trans. Pattern Anal. Mach. Intell. 23 (6) (2001) 681−685.

[25] K. Prkachin, S. Mercer, Pain expression in patients with shoulder pathology: validity, coding properties and relation to sickness impact, Pain 39 (1989) 257−265.

[26] J. Cohen, Statistical Power Analysis for the Social Sciences, Lawrence Erlbaum Associates, NJ, USA, 1988.

[27] G. Menchetti, Z. Chen, D.J. Wilkie, R. Ansari, Y. Yardimci, & A.E. Çetin. Pain Detection from Facial Videos Using Two-Stage Deep Learning. 2019 IEEE Global Conference on Signal and Information Processing (GlobalSIP), Ottawa, ON, Canada, 2019, pp. 1−5. Available from: https://doi.org/10.1109/GlobalSIP45357.2019.8969274.

[28] X. Xu, J.S. Huang, & R.De. Sa Virginia. Pain Evaluation in Video using Extended Multitask Learning from Multidimensional Measurements. In: Machine Learning for Health ML4H at NeurIPS 2019, pp. 141−154, 2020.

[29] D. Liu, F. Peng, A. Shea, R. Picard, et al. Deepfacelift: interpretable personalized models for automatic estimation of self-reported pain. arXiv preprint arXiv:1708.04670, 2017.

[30] L. Martinez, D. Rosalind Picard, et al. Personalized automatic estimation of selfreported pain intensity from facial expressions. In: IEEE Conference on Computer Vision and Pattern Recognition Workshops, pp. 70−79, 2017.

[31] R. Yang, et al. On pain assessment from facial videos using spatio-temporal local descriptors. In: 2016 Sixth International Conference on Image Processing Theory, Tools and Applications (IPTA), Oulu, Finland, 2016, pp. 1−6. Available from: https://doi.org/10.1109/IPTA.2016.7820930.

[32] L. Dai, J. Broekens, & K.P. Truong. Real-time pain detection in facial expressions for health robotics. In: 2019 8th International Conference on Affective Computing and Intelligent Interaction Workshops and Demos (ACIIW), Cambridge, United Kingdom, 2019, pp. 277−283. Available from: https://doi.org/10.1109/ACIIW.2019.8925192.

The role of social media in the battle against COVID-19

Carmela Comito

National Research Council of Italy (CNR), Institute for High Performance Computing and Networking (ICAR), Rende, Italy

6.1 Introduction

Online social media, such as Twitter and Facebook, are widely used worldwide for information dissemination. Since the onset of COVID-19, social media have been flooded with messages concerning news, updates, and, in general, information of all nature about the pandemic. This phenomenon on large scale produced huge amount of data about the COVID-19 pandemic, data that represent a precious resource for medical and government institutions to fight the pandemic. In particular, social media data could support a variety of medical and healthcare applications, like clinical trials and decision support, disease surveillance, personalized medicines, and population health management. This is made possible by the specificity and uniqueness of social media that combines textual, temporal, geographical, and network data, allowing the development of novel applications by integrating data of different natures like, for example, exploring human mobility patterns within epidemic models to monitor and predict COVID-19 spreading.

In order to exploit social media data, researchers and IT companies started to collect COVID-19 data from online social media. The majority of such efforts collected large-scale datasets and shared them publicly to enable further research. Some datasets are restricted to a single language such as English [1,2], while some others contain multiple languages [3,4].

In this context, artificial intelligence (AI) techniques can play a key role. Artificial intelligence is changing the landscape of health care and precision medicine. Social media networks, when combined with big data applications, enable the development of health applications that will result in high-quality health delivery and reduced costs. Examples of such applications include COVID-19 detection and diagnosing, population screening, positive patient monitoring, real-time tracking transmission, medical treatments, disease surveillance, and epidemiological studies.

Starting from the early days of COVID-19, several research studies exploiting social media data within AI techniques to address the epidemics-related issues have been put in place. The research activities focused on different topics: from the analysis of people reactions and the spread of COVID-19 [5–8] to the search for conspiracy theories and social activism [9,10] and the identification of misinformation propagation [11,12]. A certain number of research works are also devoted to the detection of trending topics of discussion about COVID-19 [4,5,13–15]. Another interesting branch of research concerns sentiment analysis and emotion detection of COVID-19-related social media data [16–20]. Topic modeling, when coupled with sentiment analysis, helps to find the ongoing topics being discussed on social media platforms improving sentiment understanding. Based on the user's sentiment toward the detected topics, public authorities and governments could devise the most suitable strategies. Finally, a relevant research topic investigates the effectiveness of the AI-based methods for COVID-19 forecasting and diagnosing that, in spite of the numerous models proposed by the research community, remains a challenging task in the healthcare field [21–28].

The aim of this chapter is to explore how social media data can be integrated into social and public health for pandemic preparedness. Specifically, one of the main objectives is to assess how social media data can enhance the comprehension of COVID-19 diffusion, people awareness and reactions, and how this data can be exploited to forecast COVID-19 spreading [29,30].

In this chapter, we consider an extensive collection of papers published in the last 2 years dealing with the above-cited topics of COVID-19 based on AI technologies. The aim of the study is to investigate the effectiveness of the AI-based methods proposed in combination with social media and discuss the main characteristics of some of the most promising of them. It is worth pointing out that several survey papers describing how artificial intelligence can help to better manage the pandemic have been published [31–36]. However, none of them focus specifically on the role of social media, and compared to them, the scope of this chapter is not to cover the broad range of applications proposed by exploiting AI, rather the aim is to give an overview of a subset of the relevant proposals concerned with the above-cited topics.

The chapter is organized as follows. In Section 6.2, the methodology adopted for selecting the papers is described. Section 6.3 describes some of the most significant related reviews. Section 6.4 gives an overview of the main approaches proposed in the literature and a description of the AI methods used for COVID-19 topic detection and sentiment analysis. Section 6.5 presents some of the most representative approaches devoted to COVID-19 misinformation detection and spreading over social media. The most promising AI-based proposals for COVID-19 forecasting and diagnosing are surveyed in Section 6.6. Section 6.7 outlines the main limitations of the reviewed approaches and discusses the future research directions. Section 6.8, finally, concludes the chapter.

6.2 Materials and methods

The objective of this study is to assess how data spreading over online social media can help understand COVID-19 diffusion, people concerns and reactions.

For this purpose, the chapter provides an overview of methods, algorithms, and applications proposed for forecasting and tracking COVID-19, identifying the main topics of discussion together with people feelings and emotions, as well as misinformation detection. The review of literature is carried out on databases of ScienceDirect (SD), IEEE Xplore, Web of Science (WoS), Google Scholar, Scopus, PubMed, ACM Digital Library, arXiv, and medRxiv. The search has been conducted using keywords related to the above topics under the concept of AI like *Coronavirus, artificial intelligence, machine learning, deep learning, COVID-19, forecasting, prediction, tracking, spreading, topic detection, fake news misinformation, sentiment analysis*.

A comprehensive literature search was conducted in the above-mentioned databases for English language papers published from February 2020 till now. We selected peer-reviewed articles, both journal and conference papers, and pre-prints. These articles were further screened based on title and abstract to check their compatibility with the targeted topics.

6.3 **Related reviews**

A certain number of previously published surveys focused on the different approaches proposed in the literature that are based on artificial intelligence techniques to tackle COVID-19, covering different issues like diagnosing, prediction, and screening. In this section, some of the most recent reviews are overviewed.

Wynantsm et al. [37] analyzed machine learning-based prediction models for diagnosing COVID-19 in patients with suspected infection, for prognosis of patients with COVID-19, and for detecting people in the general population at increased risk of COVID-19 infection or being admitted to hospital with the disease.

Chen et al. [31] investigated different topics concerning AI-based COVID-19 research and applications, ranging from disease detection and diagnosis to virology and pathogenesis, drug and vaccine development, epidemic and transmission prediction. The authors also highlighted the available data and resources that can be exploited for research purposes. Finally, the main challenges and potential directions of AI in fighting COVID-19 are discussed. Similar topics are discussed by Naude et al. [32] and Pham et al. [33]. Another review of machine learning and AI algorithms for managing the pandemic with respect to different application scenarios is performed in ref. [38].

In Tayarani [39], a detailed overview of the applications of AI in a variety of fields is presented, including diagnosis of the disease via different types of tests and symptoms, monitoring patients, identifying the severity of a patient, processing COVID-19-related imaging tests, pharmaceutical studies.

The survey by Hussain et al. [36] overviews existing AI techniques in clinical data analysis, including neural systems and classical support vector machine

(SVM). Alamo et al. [40] focused the analysis on data-driven methods for monitoring, modeling, and forecasting. Big data and data science approaches were instead the object of discussions of both Braghazzi et al. [41] and Latif et al. [42].

Dagliati et al. [43] surveyed collaborative data infrastructures to support COVID-19 research and applications. The paper also highlighted open issues concerning data sharing and data privacy, mainly focusing on the data interoperability problem due to the heterogeneity of data formats and standards.

Differently from other existing surveys on the subject, this paper proposes a perspective from a different point of view centered around the role that social media can play in fighting COVID-19.

Even if several research studies exploiting social media data and AI techniques to address the epidemics-related issues have been published, to the best of our knowledge there is not a comprehensive survey focusing on such topics. Therefore, the study presented in this chapter is the first exhaustive and focused survey concerning a very specific topic that is the use of social media and AI techniques to address COVID-19 issues. In fact, though there are some other interesting reviews focusing on the role of AI techniques in the battle against COVID-19, they cover a more broader spectrum of applications and topics and do not focus on the specific role of social media. Conversely, the scope of this chapter is not to cover the broad range of applications proposed by exploiting social media, rather the aim is to give an overview of some of the most representative approaches regarding the four analyzed research lines: topic detection, sentiment analysis, misinformation, and COVID-19 forecasting.

6.4 Understanding COVID-19 data

Since the early days of COVID-19, researchers started to analyze data disseminated over social media concerning the pandemic. The very first studies mainly aimed at gathering COVID-19 datasets and sharing them publicly to enable further research [1−4,44].

The collected datasets differentiate the language used in the social media posts. The majority of them are referring to the English language [1,2], but there are also other datasets concerning multiple languages [3,4]. Another distinguishing characteristic of the data collection is the temporal horizon covered. The dataset in ref. [3] is one of the long-running collections with the largest amount of tweets (i.e., 250 million), also due to its multilingual nature. Differently from the majority of the datasets, the one of [1,2] other than the raw content obtained from Twitter associates a sentiment score with each tweet. The dataset Geocov19 [44] enriches the raw tweet content with additional geolocation information.

Understanding social media data about COVID-19 played and still can play a key role in the development of novel applications supporting both medical and government institutions to fight the pandemic. In this context, topic detection and sentiment analysis techniques are the main means which analyze and comprehend people awareness and concerns about the COVID-19 pandemic. In the following section, some of the most representative approaches proposed for both research lines are overviewed.

6.4.1 Topic detection

Topic analysis (also called topic detection or topic modeling) is a machine learning technique that organizes and understands large collections of text data, by assigning "tags" or categories according to each individual text's topic or theme.

Topic analysis is a natural language processing (NLP) technique that allows to automatically extract meaning from text by finding patterns and unlock semantic structures within texts to identifying recurrent themes or topics. The two most common approaches for topic analysis with machine learning are NLP topic modeling and NLP topic classification.

In the last years, there has been a significant research effort on detecting topics and events in social media. This research line received also great attention for studying COVID-19 data on social media. Among the techniques defined for traditional data, and often adapted for Twitter data, as reported in ref. [45], the most representative method is *Latent Dirichlet Allocation (LDA)* [46]. is a topic model that relates words and documents through latent topics. It associates with each document a probability distribution over topics, which are distributions over words. A document is represented with a set of terms, which constitute the observed variables of the model. One of the main drawbacks of is that data structures used to represent the textual content of tweets are fixed in advance as the size of word vocabulary, the set of terms used, and the expected number of topics.

Direct application of traditional approaches to topic detection like LDA on Twitter streams, as pointed out in ref. [45], may give low-quality results; thus, many different methods have been proposed [19,47−49]. Generally, the common approach is to extract different data features from social media streams and then summarize such features for topic/event detection tasks by exploiting the information over content, temporal, and social dimensions. To overcome the limitations of the current state of the art, [29,50] proposed an approach for COVID-19 topic detection in Twitter that combines peak detection and clustering techniques. The approach exploits a set of spatial−temporal features of the geotagged posts [30]. Both numeric and textual features are extracted from the posts, and their temporal evolution is monitored along time to identify bursts in their values. The textual content of the tweets is represented by a set of features capturing the key elements of the text, like hashtags and words. The method proposed consists of a two-phase approach that first detects peaks in the time series associated with the spatiotemporal features of the tweets and, then, clusters the textual features (either hashtags or words) exhibiting peaks within the same timestamp. The clustering approach is based on the co-occurrence of the textual features in the tweets.

Results, performed over real-world datasets of tweets related to COVID-19 in the United States, show that the approach is able to accurately detect several relevant topics of interest. Debates range from comparisons to other viruses, health status, and symptoms, to government policy and economic crisis, while the largest volume of interaction is related to the lockdown and the other countermeasure and restrictions adopted to fight the pandemics. The main differences between the approach proposed in ref. [30] and the state of the art lie in the summarization technique. In refs. [19,47−49], the textual content of tweets is represented with a traditional vector space model, where the vector dimension is fixed in advance as the size of the word vocabulary. In streaming scenarios, because word vocabulary dynamically changes over time, it is very computationally expensive to recalibrate the inverse document frequency of TF-IDF. Differently, the approach in ref. [30] deals with content evolving signature structure of cluster by either updating frequencies of already present terms or including new terms; since we refer to term frequencies as relative values, there is no need of recalibrating the vocabulary size.

Apart from the approach presented in ref. [30], to the best of our knowledge, all the works so far proposed to detect COVID-19 topics from social media adopt the LDA approach; thus, all of them suffer from the above-cited limitations. The main relevant drawback is that the number of topics and the data structure and size has to be fixed in advance. Anyhow, in order to give a complete view of what has been done to identify topics related to COVID from social media, the current state of the art is briefly surveyed below.

In ref. [14], authors use Twitter data to explore and illustrate five different methods to analyze the topics, key terms and features, information dissemination and propagation, and network behavior during COVID-19. The authors use pattern matching and topic modeling using LDA in order to select 20 different topics about the spreading of COVID-19 cases, healthcare workers, and personal protective equipment. Using various analyses, the authors were able to detect only a few high-level topic trends. Alrazaq et al. [5] also performed topic modeling using word frequencies and LDA with the aim to identify the primary topics shared in the tweets related to COVID-19.

Chen et al. [4] analyzed the frequency of 22 different keywords such as "Coronavirus," "Corona," "Wuhan," analyzed across 50 million tweets from January 22, 2020, to March 16, 2020. Thelwall [15] also published an analysis of topics for English-language tweets during the period March 10−29, 2020. Singh et al. [11] analyzed the distribution of languages and the propogation of myths.

Sharma et al. [51] implemented sentiment modeling to understand the perception of public policy on intervention policies such as "social distancing" and "work from home." They also track topics and emerging hashtags and sentiments over countries. Again, for topic detection they used LDA.

Cinelli et al. [52] compared Twitter against other social media platforms Instagram, YouTube, Reddit, and Gab to model information spread about COVID-19. They analyzed engagement and interest in the COVID-19 topics and provide a differential assessment on the evolution of the discourse on a global scale for each platform and their users. Authors fit information spreading with epidemic models characterizing the basic reproduction numbers R0 for each social media platform. Also in this case, for topic detection a standard LDA method has been used.

In Kabir and Madria [13], CoronaVis is described, a web application allowing to track, collect, and analyze tweets related to COVID-19 generated from the United States. The tool allows to visualize topic modeling, analyze user movement information, study subjectivity, and model human emotions during the COVID-19 pandemic. That analysis is updated in real time. They also share a cleaned and processed dataset named CoronaVis Twitter dataset (focused on the United States) available to the research community at https://github.com/mykabir/COVID19. Also, in this work for topic modeling authors used LDA to find out the relevant topics.

6.4.2 Sentiment analysis

Sentiment analysis, also known as opinion mining or emotion AI, is the use of natural language processing, text analysis, computational linguistics, and biometrics to systematically identify, extract, quantify, and study affective states and subjective information.

Sentiment analysis focuses on the polarity of a text (positive, negative, neutral), but it also goes beyond polarity to detect specific feelings and emotions (angry, happy, sad, etc.), urgency (urgent, not urgent), and even intentions (interested vs. not interested). It involves the use of data mining, machine learning (ML), and artificial intelligence (AI) to mine text for sentiment and subjective information.

In the last few years, sentiment classification of social media data received great interest mostly for the specificity of microblogging posts characterized by short and noisy contents. In ref. [17], authors tested the suitability for sentiment classification of standard classifiers, like Naive Bayes, support vector machine (SVM), and maximum entropy classifiers. These classifiers were trained using emoticons in the tweets as labels, together with different textual features as words and hashtags, both unigrams and bigrams. Several approaches exploited only the hashtags for sentiment analysis detection like Davidov et al. [18]. Wang et al. [19] considered hashtags as topics, and their objective was to estimate the sentiment related to a hashtag.

Sentiment analysis and emotion detection on social media during COVID-19 gained a lot of attention from the research community due to the ability to provide epidemics information concerning both the social aspects and the overall dynamics of the population.

In Samuel et al. [16], the authors exploited only keywords related to fear to classify the sentiment of people in United States using Naïve Bayes and logistic regression as classifiers. Gao et al. [20] considered the emotions about distinctive characteristics of mental health issues, as caused by the measure taken to contrast the diffusion of COVID-19 like self-quarantining and lockdown. In ref. [53], the authors exploited Twitter to analyze people sentiments in India after the lockdown measures. The paper shows that the sentiment of people toward lockdown was positive meaning that they agreed with the government decisions.

Sharma et al. [51] analyzed the sentiments of a number of topics related to COVID-19 with the main aim of detecting misinformation spreading. The authors also raised interesting issues to address in future work in order to enhance the classification task. The work of Pedrosa et al. [54] focused on the psychological effect of COVID-19. By analyzing the nature and prevailing mood of human behavior, Pedrosa et al. found that people anxiety increased because of COVID-19 news.

As reported in ref. [55], several research works have been proposed for sentiment analysis of tweets combining different DL and NLP approaches. Sentiments toward mask usage as a preventative Mukherjee et al. [56] used a bidirectional RNN model to learn patterns of relations from textual data for sentiment analysis. Another research gathered 1 million tweets to capture people's sentiments toward mask usage as a preventative strategy in the COVID-19 pandemic [57]. They also utilized NLP to analyze the growth in the frequency of positive tweets.

In ref. [58], CovidEmo is proposed, an emotion corpus containing 3 K tweets collected during COVID-19. Authors performed a comprehensive analysis of emotional expression in CovidEmo, comparing their corpus with other datasets in the literature by experimenting with large-scale pretrained language models including BERT, TweetBERT, and COVID-Twitter-BERT. The analyses revealed the emotional toll caused by COVID-19 and the changes in the social narrative and associated emotions over time.

Zunera et al. [59] proposed an approach for sentiment classification using various feature sets and classifiers. Tweets are categorized into positive, negative, and neutral sentiment classes. The authors evaluated the performance of machine learning (ML) and deep learning (DL) classifiers using a set of evaluation metrics and a valuable feature set with the goal of improving accuracy. Experiments prove that the approach provides better accuracy compared to all other methods used in this study as well as compared to the existing approaches and traditional ML and DL algorithms.

In the work by Lwin et al. [60], COVID-19 emotion detection has been investigated by examining the worldwide Twitter trends about the pandemic, utilizing more than 20,325,929 tweets. Authors applied a lexical approach by analyzing four emotions (fear, anger, sadness, and joy) and the narratives that underlie these emotions. As an outcome of the study, they found that negative emotions were dominant during the COVID-19 pandemic.

Another interesting work about sentiment analysis of COVID-19 is proposed by Raamkumar et al. [61]. The authors aimed to understand the communication strategies that are conducted by various public health authorities to examine public sentiment and responses to COVID-19 in social media. The main objective of the study was to improve public information dissemination practices. Facebook posts and comments were utilized, and sentiment analysis was measured by using frequency, mean sentiment polarity (SP), positive-to-negative sentiment ratio (PNSR), and positive-to-negative emotion ratio (PNER).

Li et al. [62] explored the mental impact of COVID-19 on the Chinese via predictive modeling and sentiment analysis. In this regard, a text mining system was developed by the Chinese Academy of Sciences to extract content features, combined with psychoanalytic dictionary toward categorizing microblog content into linguistic annotations. The latter included various emotions, such as positive, negative, and angry emotions. People showed negative emotions (anxiety, depression, and indignation) and less positive emotions after the declaration of COVID-19. Li et al. discussed the use of lexical- and ML-based approaches in the fight against COVID-19 by utilizing natural language processing (i.e., sentiment analysis) to classify social media content into several types and choose Weibo as a study case. A total of 367,462 posts were used to extract relevant features that were applied to train ML algorithms, such as support vector machine (SVM), Naive Bayes (NB), and random forest (RF), to learn the types of unlabeled data based on labeled data. COVID-19-related information was classified into seven types of situational information, including emotional, perception, and affiliation factors. The best results were achieved by using the best-performing RF classifier.

6.5 Misinformation identification and spreading

The term misinformation has been used in the literature after the 2016 US election, as a piece of information that lacks veracity and that could mislead people. Literature makes a distinction between misinformation as unintentional and disinformation as intentional falsity. Another definition reports misinformation as either false or misleading or fabricated information, which is predominantly unintentional and honest mistake. Also, in some definitions of misinformation, the idea of intentionality is missing, meaning misinformation can be both intentional and unintentional.

Since few days after the COVID-19 outbreak, the World Health Organization declared that the pandemic is accompanied by an infodemic, that is, an information pandemic. Unfortunately, infodemic contains not only trusted and verified information but also disinformation and misinformation fostered by the widespread use of social media. Accordingly, since December 2019, social media have been overwhelmed with posts and news about COVID-19 spreading very rapidly. Spreading deceptive information may lead to anxiety, unwanted exposure to medical remedies, and tricks for digital marketing.

According to the International Fact Checking Network's (IFCN) study between January and April 2020, the fake news spread on social media can be categorized as follows: content about symptoms, causes, and cures, government documents, spread of the virus, misrepresentation of videos and photos, comments of politicians, and conspiracies that blame particular groups, countries, or communities for the spread of the virus.

The fake news spread on social media influenced a variety of people causing serious consequences in human life worldwide. For instance, in some countries people stopped consuming nonvegetarian food as fake news was spread that animals and birds could be infected with COVID-19, and consuming nonvegetarian food may spread the virus in people. This had a severe impact on the sales of nonvegetarian food. In the same way, people risked their lives due to medication misinformation. For example, a viral fake news reported that highly concentrated alcohol could disinfect the body by killing the virus.

Researchers approached the role of social media in the spreading of COVID-19 misinformation under different perspectives: information seeking and sharing behavior, cross-platform studies of misinformation [52], users' (mis)information consumption tendency. In ref. [63], a biostatistical analysis of the COVID-19 pandemic using the K-nearest neighbor (KNN) classifier is proposed. The authors collected data based on certain information from the news topics in social media through multi-document summarization. This summarization method extracted the information based on the lexical information of the topic and social patterns. The KNN classifier predicted fake news with an accuracy of 80%.

Groza [64] studied the spread of fake news on social media through the COVID ontology. The reasoning in the natural language is converted into description logic for perceiving inconsistencies among different medical sources and identifying whether the information is true or fake.

In ref. [65], FakeCovid is described, a dataset generated from multilingual news articles. In the same paper, the authors proposed a classifier based on BERT to detect fake news from the dataset generated.

In Khan et al. [66], state-of-the-art ML classifiers are employed to classify fake news regarding COVID-19. The dataset used in this work is a fusion of fake and real news collected from several social media platforms and websites. In the first step, preprocessing is performed on the dataset to remove unwanted text; then, tokenization is carried out to extract the tokens from the raw text data collected from various sources. Later, feature selection is performed to avoid the computational overhead incurred in processing all the features in the dataset. The linguistic and sentiment features are extracted for further processing. Finally, several state-of-the-art machine learning algorithms are trained to classify the COVID-19-related dataset. These algorithms are then evaluated using various metrics. The results show that random forest classifier outperforms the other classifiers with an accuracy of 88.50%.

Ng et al. [67] studied a Singapore-based Telegram group with more than 10,000 users to explore how they react to COVID-19 misinformation. The authors categorized the users' response into four types: affirm, when users accept the misinformation; denies, when users refute the misinformation; questions, when users have doubts about the misinformation; and unrelated, when users response is not relevant to the misinformation. The study shows that most of the users deny or question (45%) misinformation, while only a few users affirm it (11%).

The study by Daley et al. [68] used ML algorithms for the automatic classification of fake news on COVID-19. The features considered for the classification included the count of motion words and relativity words, prepositions in the headlines of news websites, tone expressed, and word count. They attained 79% accuracy on fivefold cross-validation using a decision tree classifier, which outperformed other state-of-the-art classifiers.

In ref. [69], how social media users respond to different types of COVID-19 misinformation is explored, and what their emotional expressions are when they interact with such misinformation. The study is based on 11,716 comments from 876 Facebook posts on five COVID-19 misinformation and addresses two relevant research questions: (1) how ignorant social media users are about misinformation? (2) How do they react to different types of misinformation? Following a quantitative content analysis method, this study produces a few findings. The results show that most of the users trust misinformation (60.88%), and fewer can deny (16.15%) or doubt (13.30%) the claims based on proper reasons. The acceptance of religious misinformation (94.72%) surpassed other types of misinformation. Most of the users react happily (34.50%) to misinformation: the users who accept misinformation are mostly happy (55.02%) because it may satisfy their expectations, and the users who distrust misinformation are mostly angry (44.05%) presuming it may cause harm to people.

6.6 COVID-19 forecasting

This section provides an overview of some of the most representative approaches proposed in the literature discussing methods and applications of statistical learning (SL), machine learning (ML), or deep learning (DL) techniques for COVID-19 forecasting in terms of new infection, deaths, and recovery.

Different statistical learning techniques have been used for COVID-19 forecasting; the methods mainly belong to two categories: regression approaches [70] and time series approaches [71]. The regressor methods used for COVID-19 forecasting are linear regression (LR), logistic regression (LoR), polynomial regression (PR), classification via regression (CR), Gaussian regression (GR), least absolute shrinkage and selection operator (LASSO), and RIDGE regression. The latter two are regression methods that use regularization techniques for obtaining more accurate predictions. The methods belonging to this category have been used mainly on time series data of reported daily COVID-19 confirmed, recovered, death, and active cases of data from all over the world. The objective was the forecasting of one or more of the above parameters, that is, predicting the daily cases of infections, and/or deaths and recovery. The majority of the proposed approaches used data from the COVID-19 data repository managed by the

Johns Hopkins University Center for Systems Science and Engineering (JHU CSSE) [72].

The literature showed that among time series approaches, the ones based on ARIMA [21−25] and Prophet [22,73] were the most effective for COVID-19 prediction. In particular, in ref. [21], the authors show that ARIMA obtains better performance than Prophet, for most of the countries analyzed.

In ref. [74], a multiple linear regression model to forecast the number of daily confirmed COVID-19 cases is proposed. The authors examine the use of phone call data and show their usefulness in forecasting daily confirmed cases. Data used in this paper comprised the number of daily COVID-19 confirmed cases and the number of daily phone calls received at the National Health Service 111 (NHS 111) in the East Midlands region of England between March 18, 2020, and October 19, 2020.

Other approaches like Wang et al. [73] proposed to combine the logistic model of population growth and the Prophet model with the aim of improving the long-term prediction capability of the time series model in order to obtain a reliable epidemic curve and trend of the epidemic.

Data used with ML methods are mainly COVID-19 patients clinical data like laboratory findings and routine bloods, patient demographics data, and surveillance data like COVID-19 cases, deaths, and recovery. In most of the proposed approaches, the addressed prediction task was aimed at forecasting COVID-19 cases, deaths and critical cases, and recovery. Similar to the statistical learning methods, DL techniques have been used mainly on time series data of reported daily confirmed, recovered, death, and active cases of COVID-19. The addressed task was the prediction of one or more of such epidemic surveillance parameters.

Many of the proposed approaches implemented different prediction models by comparing different classifiers as summarized in the following: Bayes classifier, logistic regression, lazy classifier, meta-classifier, classification via regression, rule learner, and decision tree in Arpaci et al. [26]; decision tree, extremely randomized trees, K-nearest neighbors, logistic regression, Naive Bayes, random forest, support vector machines in Brinati et al. [75].

Other approaches adopted ensemble learning models [76,77], which showed higher performance than single ML methods. In ref. [78], a model that employs supervised machine learning algorithms to identify the features predicting the COVID-19 disease with high accuracy is presented. Features examined include age, gender, observation of fever, history of travel, and clinical details such as the severity of cough and incidence of lung infection. The authors applied different machine learning algorithms and found that the XGBoost algorithm performed with the highest accuracy (85%) in predicting and selecting features detecting the COVID-19 status, independently from the age.

Gao et al. [20] presented a Mortality Risk Prediction Model of COVID-19, named MRPMC, exploiting patient's clinical data on admission able to predict death up to 20 days in advance. MRPMC is an ensemble model including logistic regression, support vector machine, gradient boosted decision tree, and neural network. To train and validate MRPMC, the authors considered 2520 COVID-19 patients with known outcomes (discharge or death) from two affiliated hospitals in China between January 27, 2020, and March 21, 2020.

In ref. [79], an ensemble learning model for COVID-19 diagnosis, named ERLX, from blood tests is proposed. The model uses three well-known diverse classifiers, extra trees, random forest, and logistic regression, which have different architectures and learning characteristics at the first level, and then combines their predictions by using a second-level extreme gradient boosting (XGBoost) classifier to achieve better performance. The ensemble model achieved very good performance with an overall accuracy of 99.88%, an AUC of 99.38%, a sensitivity of 98.72%, and a specificity of 99.99%.

A certain number of works in the literature presented a comparative analysis of deep learning methods to predict COVID-19 cases. Shastri et al. [80] proposed a nested ensemble model using long short-term memory (LSTM). The proposed Deep-LSTM ensemble model is evaluated on intensive care COVID-19 confirmed and death cases in India, and the Deep-LSTM ensemble model using convolutional and bidirectional LSTM obtains high accuracy to forecast COVID-19.

Zeroual et al. [81] presented a comparative evaluation of deep learning methods for predicting the number of new and recovered cases. The methods used are recurrent neural network (RNN), long short-term memory (LSTM), bidirectional LSTM (BiLSTM), gated recurrent units (GRUs), and variational autoencoder (VAE). Results showed a better performance of the VAE compared to the other algorithms.

In ref. [82], the authors developed an application to predict COVID-19 by exploiting laboratory findings and using six different deep learning models, like artificial neural network (ANN), convolutional neural networks (CNNs), long short-term memory (LSTM), recurrent neural networks (RNNs), CNN-LSTM, and CNN-RNN. The best results are observed from the LSTM deep learning model with an accuracy of 86.66%, recall of 99.42%, and AUC score of 62.50%. All the deep learning models experimented in the study showed an accuracy of over 84%.

Two machine learning-based forecasting models have been proposed in ref. [29], one expressed as a specific auto-regression problem (ARMAX-based) and the other one relying on a Bayesian approach. Both forecasting approaches integrate real data about COVID-19 and Twitter data. The two data sources have specific peculiarities and are complementary since they capture different virus incidence signals in the population. Web-based data like Twitter are early indicators of epidemics by providing real-time estimate of cases. Therefore, in ref. [29] the additional real-time information of Twitter has been exploited to improve epidemics prediction. The ARMAX model proposed in ref. [29] is modeled as an auto-regressive model with exogenous inputs, where the exogenous variables are in fact determined outside of the model processed. Web-based data provide real-time information of the epidemics. Therefore, for Twitter the most recent data are always available, even at prediction time (nowcast). The prediction of cases using official data from previous days forms the auto-regressive part of the model, while Twitter data from current and previous days are the exogenous inputs. As an alternative to the ARMAX model, in ref. [29], a forecasting model exploiting the Bayesian Learning paradigm is also proposed. Using a Bayesian approach and a prior epidemic COVID-19 history, the model in ref. [29] forecasts the total number of daily cases for the next few days. Bayesian inference combines uncertainty propagation of measured data with available prior information of model parameters.

6.7 Discussion: challenges and future directions

The aim of the chapter was to assess the role of social media in the battle against the COVID-19 pandemic. For this purpose, the chapter investigated four main research areas around social media data that received a lot of attention since the COVID-19 outbreaks: topic detection and sentiment analysis, misinformation, COVID-19 detection, and forecasting.

Of these research areas, topic detection and sentiment analyses are the ones that are more mature and provided the most meaningful results. However, in the case of topic detection the great majority of the approaches rely on the well-known LDA model. The main relevant drawback of LDA is that the number of topics should be fixed in advance, and the data structure has also to be fixed in advance. Accordingly, as already highlighted earlier, in a dynamic context like social media, where word vocabulary dynamically changes over time, LDA may yield low accurate results. Therefore, even if some proposal already addresses the problem with solutions specifically tailored to social media data (see ref. [30]), still there is a lack of proposals designed for social media streams.

For what concerns sentiment analysis, the main limitation of most of the approaches is the use of one sentiment lexicon to identify positive and negative sentiments and usually one sentiment lexicon to classify the tweets into categories such as fear, sadness, anger, and disgust. Varying information categories have the potential to influence human beliefs and decision-making, and hence, it is important to consider multiple social media platforms with differing information formats (such as short text, blogs, images, and comments) to gain a holistic perspective. At present, the majority of the approaches for sentiment analysis in literature are intended to generate rapid insights for COVID-19-related public sentiment using Twitter data, also exploring the viability of machine learning classification methods for sentiment analysis. In general, in order to understand COVID-19 data from social media, the main limitations are due to the nature of data in this context. In fact, the challenges and issues are related to the noisy nature of data from social media, the irrelevancy of data, and the lack of enough data to train the AI tools used for both topic detection and sentiment classification.

To ease the detection of misinformation, traditional ML and DL methods are widely used to develop systems able to classify misinformation. In particular, we have provided a comprehensive view of different misinformation types and discussed existing methodologies to detect COVID-19 misinformation focusing on feature extraction methods, classification, and detection performance. Compared with the existing techniques, DL methods appeared as one of the most efficient and effective techniques to classify misinformation accurately. Although sometimes the performance degrades, traditional ML methods also perform very well in the misinformation classification task. A few works implemented ensemble methods to build more complex and effective models to better utilize extracted features. In fact, ensemble methods combine several classifiers to learn a stronger one that is more robust than any individual classifier alone.

COVID-19 forecasting and detection is the last topic discussed in the chapter about the role of social media during the pandemic. The chapter overviewed the most significant artificial intelligence contributions for forecasting, detecting, and diagnosing COVID-19 infection cases. Currently, AI mainly focuses on medical image inspection, genomics, drug development, and transmission prediction. Thus, AI still has great unexplored potential mainly in terms of number of new cases and death prediction. One of the reasons why until now AI was not fully explored in tracking and predicting COVID-19 cases is the lack of a vast amount of historical data to train the AI models. Training models on unrepresentative datasets lead to poor and even misleading outcomes. This severely affected the performance and accuracy of the forecasting models. Today, the availability of COVID-19 surveillance data in terms of number of daily and cumulative cases and number of deaths is not an issue anymore. In fact, several collections of detailed data are available from different sources, for example, the one gathered by the Coronavirus Research Center of Johns Hopkins University. Another limitation to the development of effective COVID-19 prediction systems is the lack of usage of any exogenous variable in the forecasting process. Accounting restrictive measures like lockdown, quarantine, and traveling limitations could enhance the prediction accuracy. Furthermore, the availability of vaccination data could be integrated into the forecasting models, improving the performance of the prediction.

6.8 Conclusion

The chapter presented a survey of some of the most representative research efforts devoted to the exploitation of social media data to fight COVID-19. Different aspects and topics have been covered: spanning from the very early proposals mainly targeting at collecting in a structured way COVID-19 social media data, going through approaches for identifying topics of discussion and sentiment and emotion analysis. Misinformation detection has also been studied, and the main research findings have been outlined. An extensive overview of the application of AI technologies for COVID-19 forecasting and diagnosing is also provided. The study considered a collection of papers employing AI methodologies. More in detail, a comprehensive analysis of the main methods using statistical learning, machine learning, and deep learning has been made, and the results these methods obtained have been discussed. We also revealed the limitations of the existing studies and mentioned several research directions for further investigation in the future.

While the research effort in this field is significant and relevant, there is still much work to be done to improve current technologies in order to provide efficient and effective tools for understanding pandemics, people awareness and sentiments, and tools for detecting, tracking, and diagnosing COVID-19.

References

[1] R. Lamsal, Design and analysis of a large-scale COVID-19 tweets dataset, Appl Intell 51 (2021) 2790−2804. Available from: https://doi.org/10.1007/s10489-020-02029-z.

[2] R. Lamsal, Corona virus (covid-19) geolocation-based sentiment data. (2020), Available from: https://doi.org/10.21227/fpsb-jz61.

[3] J.M. Banda, R. Tekumalla, G. Wang, J. Yu, T. Liu, Y. Ding, et al., A large-scale covid-19 twitter chatter dataset for open scientific research − an international collaboration (2020). arXiv:2004.03688.

[4] E. Chen, K. Lerman, E. Ferrara, Tracking social media discourse about the covid-19 pandemic: development of a public coronavirus twitter data set, JMIR Public. Health Surveill. 6 (2) (2020) e19273.

[5] A. Abd-Alrazaq, D. Alhuwail, M. Househ, M. Hamdi, Z. Shah, Top concerns of tweeters during the covid-19 pandemic: infoveillance study, J. Med. Internet Res. 22 (4) (2020) e19016.

[6] L. Li, Q. Zhang, X. Wang, J. Zhang, T. Wang, T. Gao, et al., Characterizing the propagation of situational information in social media during covid-19 epidemic: a case study on weibo, IEEE Trans. Computational Soc. Syst. 7 (2) (2020). Available from: https://doi.org/10.1109/TCSS.2020.2980007.

[7] M.T. Rashid, D. Wang, Covidsens: a vision on reliable social sensing for covid-19 (2020). arXiv:2004.04565.

[8] L. Schild, C. Ling, J. Blackburn, G. Stringhini, Y. Zhang, S. Zannettou, "Go eat a bat, chang!": an early look on the emergence of sinophobic behavior on web communities in the face of covid-19 (2020). arXiv:2004.04046.

[9] E. Ferrara, What types of covid-19 conspiracies are populated by twitter bots? First Monday (2020). Available from: https://doi.org/10.5210/fm.v25i6.10633.

[10] S. Shahsavari, P. Holur, T.R. Tangherlini, V. Roychowdhury, Conspiracy in the time of corona: automatic detection of covid-19 conspiracy theories in social media and the news (2020). arXiv:2004.13783.

[11] L. Singh, S. Bansal, L. Bode, C. Budak, G. Chi, K. Kawintiranon, et al., A first look at covid-19 information and misinformation sharing on twitter (2020). arXiv:2003.13907.

[12] K. R, A.J. J, K. A, E.A. MB, K. B, A. E, et al., Coronavirus goes viral: quantifying the covid-19 misinformation epidemic on twitter, Cureus 12 (3) (2020).

[13] M.Y. Kabir, S. Madria, Coronavis: a real-time covid-19 tweets analyzer (2020). arXiv:2004.13932.

[14] C. Ordun, S. Purushotham, E. Raff, Exploratory analysis of covid-19 tweets using topic modeling, umap, and digraphs (2020). arXiv:2005.03082.

[15] M. Thelwall, S. Thelwall, Retweeting for covid-19: Consensus building, information sharing, dissent, and lockdown life (2020). arXiv:2004.02793.

[16] J. Samuel, G.G.M.N. Ali, M.M. Rahman, E. Esawi, Y. Samuel, Covid-19 public sentiment insights and machine learning for tweets classification, Information 11 (6) (2020) 314.

[17] B. Le, H. Nguyen, Twitter sentiment analysis using machine learning techniques, in: H.A. Le Thi, N.T. Nguyen, T.V. Do (Eds.), Advanced Computational Methods for Knowledge Engineering, Springer International Publishing, Cham, 2015, pp. 279−289.

[18] D. Davidov, O. Tsur, A. Rappoport, Enhanced sentiment learning using twitter hashtags and smileys, in: Proceedings of the 23rd International Conference on Computational Linguistics: Posters, Association for Computational Linguistics, USA, 2010, p. 241−249.

[19] M. Wang, K. Yang, X.-S. Hua, H.-J. Zhang, Towards a relevant and diverse search of social images, IEEE Trans. Multimed. 12 (8) (2010) 829−842.

[20] X. Gao, J. Cao, Q. He, J. Li, A novel method for geographical social event detection in social mediain Internet Multimed. Comput. Serv. (2013) 305−308.

[21] N. Kumar, S. Susan, COVID-19 pandemic prediction using time series forecasting models, in: 11th International Conference on Computing, Communication and Networking Technologies, ICCCNT 2020, Kharagpur, India, July 1−3, 2020, 2020, pp. 1−7.

[22] S. Singh, K.S.P.S.J.S. Makkhan, J. Kaur, S. Peshoria, J. Kumar, Study of ARIMA and least square support vector machine (LS-SVM) models for the prediction of SARS-CoV-2 confirmed cases in the most affected countries, Chaos, Solitons Fractals 139 (2020) 110086.

[23] A. Hernandez-Matamoros, H. Fujita, T. Hayashi, H. Perez-Meana, Forecasting of covid19 per regions using arima models and polynomial functions, Appl. Soft Comput. 96 (2020) 106610.

[24] F. Shahid, A. Zameer, M. Muneeb, Predictions for covid-19 with deep learning models of lstm, gru and bi-lstm, Chaos, Solitons Fractals 140 (2020) 110212.

[25] J. Devaraj, R. Madurai Elavarasan, R. Pugazhendhi, G. Shafiullah, S. Ganesan, A.K. Jeysree, et al., Forecasting of COVID-19 cases using deep learning models: Is it reliable and practically significant? Results Phys. 21 (2021) 103817.

[26] I. Arpaci, S. Huang, M. Al-Emran, M. Al-Kabi, M. Peng, Predicting the covid-19 infection with fourteen clinical features using machine learning classification algorithms, Multimed. Tools Appl. 80 (8) (2021) 11943−11957.

[27] G. Pinter, I. Felde, A. Mosavi, P. Ghamisi, R. Gloaguen, Covid-19 pandemic prediction for hungary; a hybrid machine learning approach, Mathematics 6 (2020).

[28] D. Assaf, Y. Gutman, Y. Neuman, G. Segal, S. Amit, S. Gefen-Halevi, et al., Utilization of machine-learning models to accurately predict the risk for critical covid-19, Intern. Emerg. Med. 15 (8) (2020).

[29] C. Comito, How covid-19 information spread in us the role of twitter as early indicator of epidemics, IEEE Trans. Serv. Comput. (2021) 1-1. Available from: https://doi.org/10.1109/TSC.2021.3091281.

[30] C. Comito, Covid-19 concerns in us: topic detection in twitter, in: 25th International Database Engineering and Applications Symposium, IDEAS 2021, Association for Computing Machinery, New York, NY, USA, 2021, p. 103−110. Available from: https://doi.org/10.1145/3472163.3472169.

[31] J. Chen, K. Li, Z. Zhang, K. Li, P.S. Yu, A survey on applications of artificial intelligence in fighting against COVID-19, CoRR arxiv2007.02202 (2020). arXiv:2007.02202. URL https://arxiv.org/abs/2007.02202

[32] W. Naudé, Artificial intelligence vs COVID-19: limitations, constraints and pitfalls, AI Soc. 35 (3) (2020) 761−765.

[33] Q.-V. Pham, D.C. Nguyen, T. Huynh-The, W.-J. Hwang, P.N. Pathirana, Artificial intelligence (AI) and big data for coronavirus (COVID-19) pandemic: A survey on the state-of-the-arts, IEEE Access. 8 (2020) 130820−130839.

[34] J. Bullock, A. Luccioni, K. Pham, C. Lam, M. Luengo-Oroz, Mapping the landscape of artificial intelligence applications against COVID-19, J. Artif. Intell. Res. 69 (2020) 807−845.

[35] F. Kamalov, A. Cherukuri, H. Sulieman, F.A. Thabtah, A. Hossain, Machine learning applications for COVID-19: A state-of-the-art review, CoRR arxiv2101.07824 (2021). arXiv:2101.07824. URL https://arxiv.org/abs/2101.07824.

[36] A.A. Hussain, O. Bouachir, F. Al-Turjman, M. Aloqaily, AI techniques for COVID-19, IEEE Access. 8 (2020) 128776−128795.

[37] L. Wynants, B. Van Calster, G.S. Collins, R.D. Riley, G. Heinze, E. Schuit, et al., Prediction models for diagnosis and prognosis of COVID-19: systematic review and critical appraisal, BMJ 369 (2020) m1328.

[38] S. Lalmuanawma, J. Hussain, L. Chhakchhuak, Applications of machine learning and artificial intelligence for COVID-19 (sars-cov-2) pandemic: a review, Chaos, Solitons Fractals 139 (2020) 110059.

[39] M.-H. Tayarani, Applications of artificial intelligence in battling against COVID-19: a literature reviewN Chaos, Solitons Fractals 142 (2021) 110338.

[40] T. Alamo, D.G. Reina, P. Millán, Data-driven methods to monitor, model, forecast and control covid-19 pandemic: leveraging data science, epidemiology and control theory, arXiv:2006.01731 (June 2020).

[41] N.L. Bragazzi, H. Dai, G. Damiani, M. Behzadifar, M. Martini, J. Wu, How big data and artificial intelligence can help better manage the COVID-19 pandemic, Int. J. Environ. Res. Public. Health 17 (9) (2020).

[42] S. Latif, M. Usman, S. Manzoor, W. Iqbal, J. Qadir, G. Tyson, et al., Leveraging data science to combat COVID-19: a comprehensive review, IEEE Trans. Artif. Intell. 1 (1) (2020) 85−103.

[43] A. Dagliati, A. Malovini, V. Tibollo, R. Bellazzi, Health informatics and EHR to support clinical research in the COVID-19 pandemic: an overview, Brief. Bioinform 22 (2) (2021) 812−822. Available from: https://doi.org/10.1093/bib/bbaa418.

[44] U. Qazi, M. Imran, F. Ofli, Geocov19: a dataset of hundreds of millions of multilingual covid-19 tweets with location information (2020). arXiv:2005.11177.

[45] F. Atefeh, W. Khreich, A survey of techniques for event detection in twitter, Comput. Intell. 31 (1) (2015) 132−164.

[46] D.M. Blei, A.Y. Ng, M.I. Jordan, J. Lafferty, Latent dirichlet allocation, J. Mach. Learn. Res. 3 (2003). 2003.

[47] J. Yin, A. Lampert, M.A. Cameron, B. Robinson, R. Power, Using social media to enhance emergency situation awareness, IEEE Intell. Syst. 27 (6) (2012) 52−59.

[48] H. Becker, M. Naaman, L. Gravano, Beyond trending topics: real-world event identification on twitter, in: Proceedings of the Fifth International Conference on Weblogs and Social Media, 2011.

[49] L.M. Aiello, G. Petkos, C. Martin, D. Corney, S. Papadopoulos, R. Skraba, et al., Sensing trending topics in twitter, IEEE Trans. Multimed. 15 (6) (2013) 1268−1282.

[50] C. Comito, D. Falcone, D. Talia, A peak detection method to uncover events from social media, in: 2017 IEEE International Conference on Data Science and Advanced Analytics, DSAA, IEEE, 2017, pp. 459−467.

[51] K. Sharma, S. Seo, C. Meng, S. Rambhatla, Y. Liu, Covid-19 on social media: analyzing misinformation in twitter conversations (2020). arXiv:2003.12309.

[52] M. Cinelli, W. Quattrociocchi, A. Galeazzi, C.M. Valensise, E. Brugnoli, A.L. Schmidt, **et al**., The covid-19 social media infodemic (2020). arXiv:2003.05004.

[53] G. Barkur, Vibha, G. Kamath, Sentiment analysis of nationwide lockdown due to covid 19 outbreak: evidence from india, Asian J. Psychiatry 51 (2020).

[54] A.L. Pedrosa, L. Bitencourt, A.C.F. Fróes, M.L.B. Cazumbá, R.G.B. Campos, S.B.C. S. de Brito, et al., Emotional, behavioral, and psychological impact of the covid-19 pandemic, Front. Psychol. 11 (2020).

[55] L. Zhang, S. Wang, B. Liu, Deep learning for sentiment analysis: a survey, Wiley Interdiscip. Reviews: Data Min. Knowl. Discovery 8 (4) (2018) e1253.

[56] S. Mukherjee, A. Malu, B. Ar, P. Bhattacharyya, Twisent: a multistage system for analyzing sentiment in twitter, in: Proceedings of the 21st ACM international Conference on Information and Knowledge Management, 2012, pp. 2531−2534.

[57] A.C. Sanders, R.C. White, L.S. Severson, R. Ma, R. McQueen, H.C. A. Paulo, et al., Unmasking the conversation on masks: Natural language processing for topical sentiment analysis of covid-19 twitter discourse, in: AMIA Annual Symposium Proceedings, Vol. 2021, American Medical Informatics Association, 2021, p. 555.

[58] T. Sosea, C. Pham, A. Tekle, C. Caragea, J.J. Li, Emotion analysis and detection during covid-19 (2021). Available from: https://doi.org/10.48550/ARXIV.2107.11020.

[59] Z. Jalil, A. Abbasi, A.R. Javed, M. Badruddin Khan, M.H. Abul Hasanat, K.M. Malik, et al., Covid-19 related sentiment analysis using state-of-the-art machine learning and deep learning techniques, Front. Public. Health 9 (2022).

[60] M.O. Lwin, J. Lu, A. Sheldenkar, P.J. Schulz, W. Shin, R. Gupta, et al., Global sentiments surrounding the covid-19 pandemic on twitter: Analysis of twitter trends, JMIR Public. Health Surveill. 6 (2) (2020) e19447.

[61] A. Sesagiri Raamkumar, S.G. Tan, H.L. Wee, Measuring the outreach efforts of public health authorities and the public response on facebook during the covid-19 pandemic in early 2020: cCross-country comparison, J. Med. Internet Res. 22 (5) (2020) e19334.

[62] Z. Li, Y. Zheng, J. Xin, G. Zhou, A recurrent neural network and differential equation based spatiotemporal infectious disease model with application to COVID-19 (2020). arXiv:https://www.medrxiv.org/content/early/2020/07/22/2020.07.20.20158568.full.pdf.

[63] S. Bandyopadhyay, S.; Dutta, Analysis of fake news in social medias for four months during lockdown in covid-19 (2020).

[64] A. Groza, Detecting fake news for the new coronavirus by reasoning on the covid-19 ontology (2020).

[65] D. Nandini, G.K. Shahi, Fakecovid-a multilingual cross-domain fact check news dataset for covid-19 (2020).

[66] F.M. Khan, R. Gupta, Arima and nar based prediction model for time series analysis of COVID-19 cases in india, J. Saf. Sci. Resil. 1 (1) (2020) 12−18.

[67] L.H.X. Ng, J.Y. Loke, Analyzing public opinion and misinformation in a covid-19 telegram group chat, IEEE Internet Comput. 25 (2) (2021) 84−91. Available from: https://doi.org/10.1109/MIC.2020.3040516.

[68] S. Spezzano, B.P. Daley, Leveraging machine learning for automatically classifying fake news in the covid-19 outbreak. (2020).

[69] M.S. Al-Zaman, Social media and covid-19 misinformation: how ignorant facebook users are? Heliyon 7 (5) (2021) e07144.

[70] D.A. Freedman, Statistical Models. Theory and Practice, Cambridge University Press, 2005.

[71] G.E.P. Box, G.M. Jenkins, G.C. Reinsel, G.M. Ljung, Time Series Analysis, Forecasting and Control, 5th edition, Wiley, 2015.

[72] COVID-19 data repository by the Center for Systems Science and Engineering (CSSE) at Johns Hopkins University, Github Inc. Covid-19 cases. https://github.com/cssegisanddata/covid-19 (2020).

[73] P. Wang, X. Zheng, J. Li, B. Zhu, Prediction of epidemic trends in covid-19 with logistic model and machine learning technics, Chaos, Solitons Fractals 139 (2020) 110058.

[74] B. Rostami-Tabar, J.F. Rendon-Sanchez, Forecasting COVID-19 daily cases using phone call data, Appl. Soft Comput. 100 (2021) 106932.

[75] D. Brinati, A. Campagner, D. Ferrari, M. Locatelli, G. Banfi, F. Cabitza, Detection of covid-19 infection from routine blood exams with machine learning: a feasibility study, J. Med. Syst. 135 (44) (2020).

[76] L. Breiman, Bagging predictors, Mach. Learn. 24 (2) (1996) 123−140.

[77] R.E. Schapire, Boosting a weak learning by maiority, Inf. Computation 121 (2) (1996) 256−285.

[78] M.M. Ahamad, S. Aktar, M. Rashed-Al-Mahfuz, S. Uddin, P. Liò, H. Xu, et al., A machine learning model to identify early stage symptoms of SARS-Cov-2 infected patients, Expert. Syst. Appl. 160 (2020) 113661.

[79] M. AlJame, I. Ahmad, A. Imtiaz, A. Mohammed, Ensemble learning model for diagnosing COVID-19 from routine blood tests, Inform. Med. Unlocked 21 (2020) 100449.

[80] S. Shastri, K. Singh, S. Kumar, P. Kour, V. Mansotra, Deep-LSTM ensemble framework to forecast COVID-19: an insight to the global pandemic, Int. J. Inf. Technol. (Singap.) (2021).

[81] A. Zeroual, F. Harrou, A. Dairi, Y. Sun, Deep learning methods for forecasting covid-19 time-series data: a comparative study, Chaos, Solitons Fractals 140 (2020) 110121.

[82] T. Alakus, I. Turkoglu, Comparison of deep learning approaches to predict covid-19 infection, Chaos, Solitons Fractals 140 (2020). Nov.

De-identification techniques to preserve privacy in medical records

7

Rosario Catelli and Massimo Esposito

Institute for High Performance Computing and Networking (ICAR), National Research Council (CNR), Naples, Italy

7.1 Introduction

The outbreak of the COVID-19 pandemic has made the need to exploit the information contained within electronic health records increasingly stringent and vital in order to manage critical and life-threatening health situations in the best way possible. Despite the increasing availability of such data made available by the medical community, the privacy regulations in place in several countries have set clear stakes for the use of such personal data. The best-known examples are the Health Insurance Portability and Accountability Act (HIPAA) and the General Data Protection Regulation (GDPR) in effect in the United States and the European Union, respectively. While in the case of the more recent GDPR, there is still a lack of a specific procedure for automating the processes, HIPAA, on the other hand, makes available both a manual procedure called expert determination and an automated method called Safe Harbor: the latter allows through a list of 18 relevant identifiers (so-called PHI, an acronym for protected health information) to precisely identify the elements that need to be removed and/or replaced with plausible and realistic surrogates. In this regard, in recent years the debate in the scientific community has also begun to focus on the so-called de-identification phase that immediately precedes the anonymization phase of the documents: not only making them available to the general public, but also preserving their readability in order to allow the exploitation of such documentation for both health and research purposes has become an important point to work on [1].

From a technological point of view, recent years have seen a gradual shift from rule-based systems to machine and deep learning-based systems, exploiting natural language processing (NLP) techniques such as named entity recognition (NER): particularly with the latter technique, it is possible to assimilate PHI to entities to be recognized for the purpose of de-identification than anonymization. Initially, handcrafted rules were exploited to identify entities of interest, but over time the limitations of this approach were shown to be incapable of keeping pace with changing context and language. So there was a shift to classifiers based on

Artificial Intelligence in Healthcare and COVID-19. DOI: https://doi.org/10.1016/B978-0-323-90531-2.00007-2

machine learning algorithms, which required large data sets and time for hand-made feature engineering [2,3]. With deep learning, there remained the problem of needing large data sets correctly labeled for the intended purpose [4]. Newer language models rely on so-called embeddings, a dense vector representation of words, in order to exploit prior knowledge that can then be adapted to the specific task. The static nature of such representations initially posed a problem for handling so-called out-of-vocabulary words and polysemy, although the subsequent advancement toward models based on characters or subwords and capable of exploiting contextuality, looking at both what precedes and what follows in texts, has further improved performance in recognizing semantics, morphosyntactic variations typical of handwritten text, and handling out-of-vocabulary words and context-dependent polysemy: not surprisingly, as already mentioned, the scientific community has gradually realized the importance of preserving the readability of texts while preserving anonymity.

In fact, a de-identification system capable of context analysis is certainly one that can potentially aspire to better performance in the anonymization phase due to a more refined ability to classify the entities involved. To date, due to the greater amount of electronic medical records available, the English language has been the most widely used in global competitions (i2b2) although recently there have been other experiences in French (ShARe/CLEF eHealth Evaluation Lab) and Spanish (IberLEF 2019-Medical Document Anonymization track known as MEDDOCAN track). A direct consequence of this has been the limited performance of notoriously data-hungry deep learning systems, as well as the general lack of experimentation with so-called low-resource languages.

The effort of this work is to summarize and compare the efforts made in the field of clinical de-identification in the Italian language. Specifically, referring to some previous works of the authors, a performance comparison is proposed between different deep neural network architectures on the data set for clinical de-identification in the Italian language known as the SIRM COVID-19 de-identification corpus. Downstream of this comparison, the advantages and disadvantages of the different systems set up to deal with different scenarios are tried to be highlighted, and some possible future directions of development are proposed.

The Italian language, although possessing an alphabet similar to that of English, has a wide syntactic and morphological variety. For these reasons, comparing different systems, whether based on static or contextual embeddings and working at the character, subword, or word level, provides additional clues as to how to deal with such language. Just as it is possible to gain insight into what might be satisfactory combinations of embeddings capable of holding context, polysemy, and morphosyntactic variations, it is important to remember how an electronic medical record, precisely because it is often written by one or more doctors, is a particularly relevant usage scenario in that sense and, in the case of COVID-19, constitutes even more of a test bed where terminology may not yet be entirely definitive because it is a frontier scenario. Finally, the results shown employing techniques of knowledge transfer between different languages, in the specific case from English to Italian, identify directions to follow in approaches to deal with low language resource scenarios, even more so when related to factors such as the pandemic scenario where there is not only a prior problem of unavailability of language data sets but also a lack of context-specific data sets, regardless of language.

Application of the most fitting of the methods described would allow for better de-identification of Italian medical records, including those related to COVID-19, and accelerate their public dissemination, increasing institutional trust due to greater respect for privacy.

The organization of this work is as follows: in the next section, an overview of the context of the scientific literature in which this paper moves is given; instead in Section 7.3, relevant information is provided regarding the data sets and network architectures employed; and in Section 7.4, the different results are compared and discussed, before arriving at the final section in which conclusions are outlined and future directions suggested.

7.2 Background

The recurring definition in the literature for named entity [5] is assimilated to what is the concept of PHI within EHRs and, consequently, can be identified by NLP NER techniques speaking specifically of clinical NER when working on EHRs: the advantages are obvious in that it becomes possible to make content publicly available for medical investigations and research purposes while safeguarding the privacy of patients following the subsequent anonymization process that replaces PHI identified through NER with appropriate surrogates, which are well specified if the system is also able to correctly classify what is identified.

In addition to manual de-identification approaches, which suffer from the problems pointed out by Ref. [6], namely, (1) the inability to crowdsource due to limited access to data, (2) human error, and (3) cost, nowadays systems that automate the process are increasingly being developed. From a historical perspective, one discriminator in the development of such systems is the use of deep learning techniques, which is different from anything that has been developed previously [2,3].

For these reasons, the sequel is organized in order to delve into systems based on deep learning techniques in Section 7.2.1, language models and embeddings in Section 7.2.2, and finally provide, in Section 7.2.3, a broad view regarding the related concepts of clinical de-identification, low-resource languages, and transfer learning.

7.2.1 Deep learning systems

The dramatic increase in the availability of data, albeit mostly in English, has made it possible for researchers to develop named entity recognition (NER) systems based on deep learning techniques and algorithms [7−10] that were later also employed in the clinical field [6,11].

The backbones of such NER systems are basically two: one is the first layer, the input layer, generally called embedding [12] that constitutes the numerical method of representation for words (or subwords or characters), and the other is the structure of the neural network itself, which basically exploits recurrent neural networks (RNNs) [13,14] or rather their evolution, namely long short-term memory (LSTM) networks.

A NER problem, being able to be approached as a sequence labeling problem, sees in the BiLSTM architecture [7,10] and character-level representations [8,10] great allies, which even in the clinical domain immediately showed great advantages [15−18] as well as in de-identification [11,19].

Unlike in other domains, newer architectures based on transformers [20] such as BERT [21] have not shown great advantages for NER. Although domain versions such as Clinical BERT [22] or BioBERT [23] proved superior to generic versions for medical applications, and the same was not true for de-identification tasks. Even the most recent MEDDOCAN competition showed how BiLSTM networks with CRF perform particularly well compared to BERT-based systems, especially in low-resource scenarios such as the Spanish language [24]. A combination of BiLSTM with CRF leveraging the embedding layer of BERT also showed state-of-the-art results in the English language [25].

7.2.2 Language models and embeddings

As mentioned earlier, embeddings are nothing more than numerical, usually vector representations of variables such as words, subwords (which do not exactly correspond to syllables), characters, or even whole sentences. A ready-to-use embedding layer is achieved by pretraining on large corpora from a purpose-built neural network, so that this prior knowledge is then exploited for a specific task on target data that are generally small in size.

From a technological point of view, embeddings were initially static with respect to context: that is, their numerical representation did not change as the surrounding words changed, such as in the case of Word2Vec by Ref. [12] and GloVe by Ref. [26]. With the emergence of FasText by Ref. [27], there is a first attempt to go further, no longer associating embedding with a word but with subwords, i.e., sets of characters that constitute n-grams, so as to then reconstruct the embedding associated with the whole word by looking at its component subwords.

The dynamism of these representations goes hand in hand with the beginning of exploiting two identical architectural components (instead of one) for their construction: one component devoted to working on the sequence, as before, and another instead devoted to working on the reverse sequence. Altogether, such an operation makes it possible to capture the relationships between words within texts: this is the origin of the so-called statistical linguistic models, language models (LM) for short, such as ELMo [28], Flair [29], BERT [21], and GPT [30].

Interestingly, some previous approaches have been somewhat reworked at this stage: the BERT tokenizer, for example, based on the WordPieceModel [31] segmenter, always works at the subword level with an approach similar to FastText. In contrast, with Flair [29], a contextual approach that was at the character level (character level LM) was preferred, unlike ELMo built earlier but at the token level.

Language modeling is still an open problem today, and there are many studies in this regard that show, for example, how the level of learnable information is affected by the depth of the network [32] or other architectural limitations [33].

7.2.3 Clinical de-identification, low-resource languages, and transfer learning

Despite the lack of resources, de-identification and anonymization systems for languages other than English have developed greatly in recent years: Danish [34], Dutch [35,36], French [37,38], German [39,40], Norwegian [41], Polish [42,43], Portuguese [44], and Swedish [45]. Recently, Spanish language challenges have also been organized: the most recent MEDDOCAN track [24] within IberLEF 2019.[1] As far as we know, to date there is nothing similar in the Italian language, probably due to a lack of language resources: for this reason, the use of transfer learning techniques could be a solution.

In fact, one of the branches of transfer learning, and of particular interest when applied to the NLP domain, is so-called cross-language transfer learning: Ref. [46] states that it is a type of transductive transfer learning in which, although the source domain (training set) and the target domain (test set) are different (different languages in NLP), knowledge is transferred through the use of a single common representation space.

Recently, multilingual embeddings employed in different NLP tasks have appeared [47–49], but among all of them it is good to count the Byte-Pair Embeddings (BPEmb) of Ref. [50] created to address the problem of out-of-vocabulary words by exploiting subword-level mechanisms and based on the Byte-Pair Encoding (BPE) of Ref. [51]. They exist in 275 languages and have been used in different interlanguage scenarios [52–54], and there is a multilingual version (MultiBPEmb).

The contextuality of language models such as Embeddings from Language Models (ELMo) [28] and Flair [55] has been shown to be a fortiori more valid in interlanguage scenarios than static models. Ref. [56] tested a polyglot version of ELMo, with important results: in detail, the use of a single model capable of handling all languages, as opposed to language-specific models, was shown to be a

---

[1] https://sites.google.com/view/iberlef-2019

better choice, especially in low-resource scenarios. Indeed, although languages are different, they are often related to each other in some aspects such as semantics, morphology, and syntax, as also shown by Ref. [57] in the case of Slavic languages.

As mentioned above, BERT [21] is a deep contextual language model proposed by Google, based on transformers [20] and trained on Cloze Task [58], commonly known as masked language modeling, which differs from classical right-to-left or left-to-right language modeling by allowing information to be freely encoded from both directions at each level. Again, in addition to language-specific versions, there was released a multilingual version (mBERT)[2] trained on 104 languages and still much studied today [59−61]: although excellent in many scenarios, Ref. [62] showed that BPEmb is often superior in low-resource scenarios.

7.3 Material and methods

Next, Section 7.3.1 provides information on which data sets were used, while Section 7.3.2 outlines the characteristics of the architectures employed. Section 7.3.3 provides an overview of the experimental setups adopted, and Section 7.3.4 indicates how systems were evaluated. Finally, Section 7.3.5 illustrates the training strategies adopted to also leverage transfer learning.

7.3.1 Data sets

Hereafter, data sets used are presented: both use the IOB tagging format [63] to distinguish between tokens and entities. In detail, the SIRM COVID-19 de-identification corpus data set is in the Italian language, while the i2b2/UTHealth 2014 de-identification corpus data set is an English data set.

7.3.1.1 The SIRM COVID-19 de-identification corpus

The Italian SIRM COVID-19 data set, based on a collection of 115 unannotated medical records in pdf format released by SIRM,[3] was created following annotations guidelines adopted by Ref. [64] and released to the scientific community by the authors in previous work Ref. [65]. The authors used 65 medical records for training and 50 medical records for testing. PHI distribution is reported in Table 7.1.

---

[2] https://github.com/google-research/bert/blob/master/multilingual.md
[3] https://www.sirm.org/category/senza-categoria/covid-19/

Table 7.1 PHI distributions in SIRM COVID-19 de-identification corpus.

PHI category: subcategory	Training	Test	Total
AGE	63	55	118
CONTACT: PHONE	3	7	10
CONTACT: URL	66	76	142
DATE	64	90	154
ID: ID NUMBER	137	129	266
LOCATION: CITY	38	63	101
LOCATION: COUNTRY	1	5	6
LOCATION: HOSPITAL	134	132	266
LOCATION: ORGANIZATION	4	9	13
LOCATION: OTHER	3	6	9
NAME: DOCTOR	303	430	733
NAME: PATIENT	3	0	3
PROFESSION	38	27	65
PHI Category	**Training**	**Test**	**Total**
AGE	63	55	118
CONTACT	69	83	152
DATE	64	90	154
ID	137	129	266
LOCATION	180	215	395
NAME	306	430	736
PROFESSION	38	27	65
Total # of entities	**857**	**1029**	**1886**

After manual annotation in brat standoff format, the data were converted via python script from pdf file to CONLL format by adapting NeuroNER [19] and using spacy as tokenizer.

7.3.1.2 The i2b2/UTHealth 2014 de-identification corpus

It was released by Ref. [3], from the i2b2 National Center for Biomedical Computing for the NLP Shared Tasks Challenges, whose de-identification guidelines reported by Ref. [64] conform to the HIPAA Safe Harbor criteria, and Table 7.2 presents an exhaustive list of PHI distribution. It was created starting from 1304 medical records of 296 patients (2–5 records each), divided between a training data set (790 XML documents, 269 for validation) and a testing data set (514 XML documents), where named entities are annotated as text spans with corresponding entity types. The initial 18 categories of PHI identifiers given by HIPAA Safe Harbor criteria have been expanded to be more specific and hence regrouped into 7 main categories and some subcategories. To be concise, some subcategories with 0 elements were omitted from Table 7.2 because they were obviously irrelevant for de-identification purposes.

Table 7.2 PHI distribution in i2b2/UTHealth 2014 de-identification corpus.

PHI category: subcategory	Training	Validation	Test	Total
AGE	810	423	764	1997
CONTACT: EMAIL	3	1	1	5
CONTACT: FAX	5	3	2	10
CONTACT: PHONE	229	80	215	524
CONTACT: URL	2	0	0	2
DATE	5254	2248	4980	12482
ID: BIO ID	1	0	0	1
ID: DEVICE	7	0	8	15
ID: HEALTH PLAN	1	0	0	1
ID: ID NUMBER	171	90	195	456
ID: MEDICAL RECORD	398	213	422	1033
LOCATION: CITY	259	135	260	654
LOCATION: COUNTRY	53	13	117	183
LOCATION: HOSPITAL	928	509	875	2312
LOCATION: ORGANIZATION	85	39	82	206
LOCATION: OTHER	4	0	13	17
LOCATION: STATE	221	93	190	504
LOCATION: STREET	144	72	136	352
LOCATION: ZIP CODE	139	73	140	352
NAME: DOCTOR	1932	953	1912	4797
NAME: PATIENT	879	437	879	2195
NAME: USERNAME	219	45	92	356
PROFESSION	149	85	179	413
Total # of entities	11,893	5512	11,462	28,867

7.3.2 System architectures

This section illustrates which state-of-the-art system architectures have been used. In detail, the system architecture based on the BiLSTM plus CRF is described in Section 7.3.2.1, while the system architecture based on the BERT neural network is described in Section 7.3.2.2.

7.3.2.1 BiLSTM plus CRF-based architecture

One of the best-performing sequence labeling architectures recognized by scientific literature is represented by the BiLSTM plus CRF model, as stated by Ref. [7]. The combination of BiLSTM and CRF makes it possible to exploit two winning features of both layers: in the first case input features and in the second case sentence-level tags. Both layers take advantage of bidirectionality, i.e., the ability to handle information present both before and after the analyzed token so as to

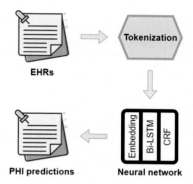

FIGURE 7.1

Overview of the clinical de-identification process through NER.

also see long-range dependencies and obtain more robust results than using individual layers, as shown by Ref. [7].

Fig. 7.1 shows an overview of the processes required for NER-based clinical
de-identification. First of all, a collection of clinical records documents is needed,
which have to be reorganized in the form of an annotated data set. Second, there
is a preliminary step to prepare the input data, i.e., tokenization: it is performed
in order to split each input sentence $s = w_1 w_2 \ldots w_n$ (where w_t, with $1 \leq t \leq n$,
represents the generic token) within raw clinical notes into a sequence of tokens.
At this point, the real neural network comes into play, whose first input layer is
called the embedding layer: here, the tokens (which can be words, subwords, or
characters, depending on the type of embedding chosen) are transformed into
numeric vectors that, in the more sophisticated versions, try to incorporate different aspects of the tokens (grammatical, morphological, syntactic, semantic, and so
on) from a general point of view. Then, the BiLSTM specializes in these representations by exploiting the ability to analyze the bidirectional context (taking
into account its memory limits), i.e., the correlations between the tokens according to the texts in which they are placed. Finally, the CRF layer, working in an
analogous way to the BiLSTM layer, takes care of providing the predictions of
the output labels, trying to preserve their coherence: for instance, it is unlikely
that the sequel of a token of type "beginning of person name" is a token of type
"continuous of thing name". Finally, a high-level algorithmic representation of the
proposed NER-based clinical de-identification management method is illustrated
with Algorithm 7.1.

7.3.2.1.1 Embedding layer

The embedding layer was chosen on a case-by-case basis according to the
requirements to be met, selecting and mixing different types of embeddings. It is
important to remember that, as shown by Ref. [22], the use of specific

Result: Most likely PHI label for each token
Given pre-processed EHRs;
while EHRs not finished do
    Tokenize documents;
    if Tokenization is successful then
        Embedding tokens;
        Specializing embeddings through context analysis given by
          BiLSTM;
        Providing probabilities of each PHI label for each token given
          by CRF;
    else
        Change EHRs pre-processing steps;
    end
end

ALGORITHM 7.1

Algorithmic representation of the proposed method.

embeddings for the clinical de-identification task, i.e., clinical or biomedical versions, does not provide improvements, so versions of embeddings trained on generic domains have been used. Embeddings used are FastText [27] and Flair [29] (also stacked together), or BPEmb and MultiBPEmb [50] stacked with Flair.

7.3.2.2 BERT-based architecture

BERT [21] is a general-purpose language model trained on a large text corpus (like Wikipedia), which can be used for various downstream NLP tasks, such as NER, relation extraction, and question answering, without heavy task-specific feature engineering. In detail, BERTBASE architecture is based on 12 encoder layers, called transformers blocks, 12 attention heads or self-attention [20], and feed-forward networks with a hidden size of 768. Instead, BERTLARGE is based on 24 encoder layers, 16 attention heads, and feed-forward networks with a hidden size of 1024. For simplicity, if not specified, we will refer to BERTBASE in the following. A simple network topology is shown in Fig. 7.2.

BERT accepts embedding and encoder input/output vectors that have a dimension of 512, called maximum sequence length. Some special tokens are employed: the first is *[SEP]*, used for segment separation. The second one corresponds to the first input token supplied, the *[CLS]* token (*CLS* stands for *Classification*), which produces an output vector, of *hidden size* dimension, that can be used as the input for an arbitrarily chosen classifier. In particular, for NER tasks, BERT is fine-tuned following a general tagging task approach without a CRF layer as the output layer. As input to the token-level classifier, working over the NER label set, the representation of the first sub-token is used.

Formally, the final hidden representation h_i of each token i is passed into softmax function. The probability P is calculated as follows:

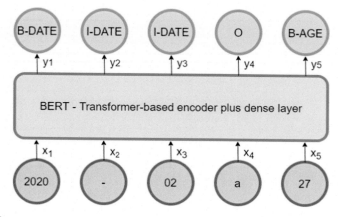

FIGURE 7.2

Simplified BERT neural network topology for NER task.

$$P(t|h_i) = softmax(W_o H_i + b_o) \tag{7.1}$$

where $t \in T$, W_o, and b_o are weight parameters. Furthermore, during the training, categorical cross-entropy as loss function is used.

7.3.3 Experimental setups

This section presents the experimental setups, all performed on an IBM POWER9 cluster with NVIDIA V100 GPUs. Each experiment was run five times, and then, the results were averaged (rounded to the fourth decimal place). All experiments were tested on ad hoc created data set, the Italian SIRM COVID-19 de-identification corpus.

7.3.3.1 BiLSTM plus CRF-based systems

Part of the experiments was based on a BiLSTM plus CRF-type architecture, to which the embedding layer was changed. The architecture overview of the proposed clinical de-identification system is shown in Fig. 7.3.

These experiments used the Flair framework [55] for BiLSTM plus CRF model implementation. It provides state-of-the-art general-purpose architectures with thousands of pretrained models in over a hundred languages for NLP tasks, such as NER, part-of-speech (PoS) tagging, sense disambiguation, and classification. Flair framework was used with the hyperparameters reported in Table 7.3, and the stochastic gradient descent (SGD) algorithm was used to estimate neural network parameters. Embeddings configurations tested are the following:

- FastText;
- Flair (forward and backward);
- FastText plus Flair (forward and backward);

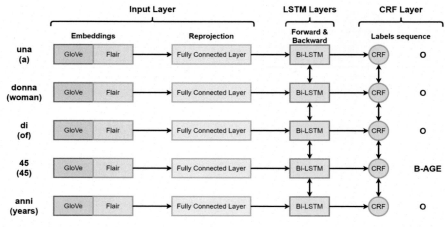

FIGURE 7.3

Overview of the BiLSTM plus CRF-type architecture.

Table 7.3 BiLSTM plus CRF architecture hyperparameters.

Hyperparameter	Value
Annealing factor	0.5
Batch size	16
Dropout (variational)	0.5
Dropout (word)	0.05
Epochs	up to 500
Gradient clipping	5
Hidden size	256
Learning rate	from 0.1 up to 0.0001
Patience (early stopping parameter)	3
RNN Layers	1

- BPEmb plus Flair (forward and backward);
- MultiBPEmb plus Flair multi-fast (forward and backward).

7.3.3.2 BERT-based systems

Besides the BiLSTM plus CRF model, the BERT model was tested too, which is another common state-of-the-art language model for different NLP tasks, as introduced in Section 7.3.2.2. In detail, the Hugging Face Transformers[4] framework

---

[4] https://github.com/huggingface/transformers

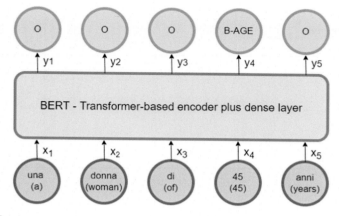

FIGURE 7.4

Overview of the BERT-type architecture.

for BERT-based models was used, whose employed architecture is shown in Fig. 7.4.

In particular, the Italian BERT models used are those made available by the MDZ Digital Library team at the Bavarian State Library.[5] The Italian BERT was trained on a source of data made by a recent Wikipedia dump and various texts from the OPUS corpora[6] collection with a final corpus size equal to about 13 GB and more than 2 billion tokens. Hyperparameters used are shown in Table 7.4, while configurations tested are the following:

- $BERT_{BASE}$ Cased;
- $BERT_{BASE}$ Uncased;
- mBERT Cased.

7.3.4 Evaluation metrics

To assess the performance of the models and compare them, the F_1 measure was used, defined as the harmonic mean of precision P and recall R. Defined TP as the number of true positives, FP the number of false positives, and FN the number of false negatives, these metrics can be defined as:

$$F_1 = \frac{2 * P * R}{P + R} \tag{7.2}$$

$$P = \frac{TP}{TP + FP} = \frac{\text{\# of correctly predicted items}}{\text{\# of predicted items}} \tag{7.3}$$

[5] https://huggingface.co/dbmdz/
[6] http://opus.nlpl.eu/

Table 7.4 BERT architecture hyperparameters.

Hyperparameter	Value
Attention heads	12
Batch size	32
Epochs	5
Hidden size	768
Languages (mBERT only)	104
Hidden layers	12
Maximum Sequence length	512
Parameters	110 M

$$R = \frac{TP}{TP + FN} = \frac{\text{\# of correctly predicted items}}{\text{\# expected items}} \tag{7.4}$$

where items are entities or tokens, evaluated in a binary way (recognized or not) or a finer way (right or wrong recognition by i2b2 category or by i2b2 subcategory).

7.3.5 Training strategies

The English i2b2 2014 de-identification corpus was used in conjunction with the Italian SIRM COVID-19 de-identification corpus to test if performances in the low-resource language scenario under study, i.e., the Italian language scenario, could be increased by leveraging different transfer learning approaches. In detail, while results were obtained on the test set of the Italian SIRM COVID-19 de-identification corpus, the training phase was different, distinguishing among four strategies:

- EN: training was done leveraging only the English data set.
- IT: training was done leveraging only the Italian data set.
- MIX: training was done by providing a mixed training set, merging the English and the Italian ones.
- EN-IT: training was done in two steps, providing first the English data set and then the Italian one.

7.4 Results and discussion

The micro-averaged F_1 results are shown in Table 7.5. In particular, column *Model* indicates the trained architecture with correlated embeddings, while column *Strategy* indicates the training strategy adopted. On the other columns, as explained in Section 7.3.4, there are six evaluation criteria: S, C, and B represent,

Table 7.5 Micro-averaged F_1 results.

Model	Strategy	S_E	C_E	B_E	S_T	C_T	B_T
BiLSTM plus CRF: FastText (IT)	IT	0.7034	0.7130	0.7297	0.7821	0.8155	0.8395
BiLSTM plus CRF: Flair (IT)	IT	**0.8100**	0.8224	0.8289	0.8797	0.9045	0.9211
BiLSTM plus CRF: FastText (IT) + Flair (IT)	IT	0.8063	**0.8294**	**0.8308**	**0.8850**	**0.9116**	0.9211
BiLSTM plus CRF: BPEmb (IT) + Flair (IT)	IT	0.8110	0.8278	0.8317	0.8856	0.9115	0.9190
BiLSTM plus CRF: MultiBPEmb + Flair multi-fast	EN	0.2662	0.2948	0.3134	0.4103	0.4914	0.5797
	IT	0.7910	0.8118	0.8159	0.8826	0.9060	0.9183
	MIX	0.8371	**0.8602**	0.8618	0.8970	0.9304	0.9417
	EN-IT	**0.8391**	0.8595	0.8619	**0.9033**	0.9321	**0.9449**
BERT$_{BASE}$ (IT) Uncased	IT	0.6442	0.6667	0.6848	0.7667	0.8083	0.8796
BERT$_{BASE}$ (IT) Cased	IT	0.7553	0.7880	0.7969	0.8561	0.8979	0.9260
mBERT Cased	EN	0.4585	0.5029	0.6878	0.5498	0.6097	0.6878
	IT	0.7768	0.8207	**0.9449**	0.8923	**0.9353**	**0.9449**
	MIX	0.7696	0.8105	0.9379	0.8833	0.9245	0.9379
	EN-IT	0.7228	0.7576	0.8969	0.8241	0.8678	0.8969

respectively, *i2b2 subcategory*, *i2b2 category*, *binary*, while the subscripts E and T stand for *entity* and *token*.

In analyzing the results, the first consideration to be made that immediately jumps out is how the use of multilingual systems does not yield satisfactory performance if the training language and the target language are different; in fact, the EN strategy always delivers rather disappointing performance with these systems, generally the worst for results.

Another general consideration, but very important in the de-identification scenario, concerns the different performances that are achieved at the token level and entity level: while the former are generally higher, it is also true that they are of lesser weight if not misleading for the evaluation of these systems, given the application scope of reference. It becomes crucial to remember how the step that follows de-identification is anonymization, and the latter is all the more effective the more valid the surrogates with which the substitution of the identified elements is made. Then, it is clear both why the identification of tokens alone is no longer sufficient and why it is of paramount importance to look at the category level and not the binary level, providing the former not only the element but also the type to which it belongs and consequently making a substitution with the surrogate more effective, limiting the chances of re-identification.

Overall, therefore, the result to be considered as a guide when comparing performance across systems is that entity at the category level: the more refined subcategory level further increases the level of difficulty and always has a lower score than the category level but, since it is always a multi-class classification, no model that overturns the performance ranking by moving from the category level to the subcategory level, rather it finds confirmation of its ranking.

Looking now at the Italian language systems (denoted by (IT) in the Model column): the BiLSTM plus CRF-based systems employing Flair manage to achieve the highest performance, even when compared with the multilingual system with an IT training strategy; on the other hand, the BERT-based systems offer a different view; whereby, while the greater effectiveness of cased systems over uncased is evident, mBERT with an IT training strategy still performs better than $BERT_{BASE}$ (IT) Cased.

In detail, the BiLSTM plus CRF model with stacked embedding FastText plus Flair achieves the best performance in detection, evidently combining well both the ability of FastText to handle semantic similarity by working at the subword level and the ability of Flair to handle polysemy, context, and morphosyntactic variations. Instead, on the other hand, the supremacy of the Cased variants of BERT underscores an aspect that is important to take into account in the clinical de-identification scenario: a NER system capable of distinguishing upper and lower case is definitely preferable as the entities to be identified are often proper names, whether of things, places, or people.

Finally, what emerges clearly is that it is always convenient to exploit a multilingual system if one has the opportunity to employ transfer learning approaches that employ a resource-intensive language such as English to augment the training set. Although it is also the mBERT model that benefits, the BiLSTM plus CRF model with stacked embedding MultiBPEmb plus Flair multi-fast turns out to be superior and is definitely the most suitable model for clinical de-identification in the Italian language.

Another important aspect concerns the transfer learning approach employed: it seems clear that the use of a single joint data set is preferable in the case of the mBERT model, probably going to introduce a smaller amount of "noise" into the neural network than in the situation where two fine-tuning steps are required. Instead, in the case of the BiLSTM plus CRF model, the EN-IT approach looks superior at the category entity level but drops in all the other cases, also the subcategory entity level, which suggests opting for an EN-IT approach.

In addition, as noted by the authors previously [65−67], the ever-increasing interconnectedness of people in the scientific community thanks to technological advancement makes it more and more expected to come across English or Chinese bibliometric references in, for example, EHRs in different languages: this phenomenon can only be coped with by exploiting multilingual systems that transcend the limit of monolingualism and are potentially able to intercept even entities in languages other than the target language that could nevertheless be part of the test set albeit in a very small percentage.

7.5 Conclusion

This work summarizes the state-of-the-art techniques for de-identifying EHRs using NER techniques, with particular regard to the Italian language. Systems were tested on an Italian data set for clinical de-identification created from the COVID-19 medical records made available by SIRM. The category level classification, i.e., multi-class approach, has achieved cutting-edge results, emphasizing the importance of working at this classification level in the de-identification field: if the goal is to correctly replace entities with surrogates for anonymization purposes, the binary level cannot be considered the main evaluation method as seen in previous works.

Furthermore, four strategies were compared to pinpoint the best to apply in a low-resource language scenario like the Italian one, showing that a double training, before in English than in Italian, exploiting a BiLSTM plus CRF architecture in combination with MultiBPEmb and Flair Multilingual Fast embeddings, was the best strategy.

The real limitation of this research area remains in the size of the data sets available for clinical de-identification: it would be appropriate to increase the availability of de-identification data sets of the same size as the English i2b2 2014, so as to allow a fair comparison with monolingual systems and provide strong baselines of reference before attempting necessary low-resource case study approaches, perhaps even employing other tools such as sentiment analysis [68,69] to support what is available.

References

[1] V. Vincze, R. Farkas, De-identification in natural language processing, 37th International Convention on Information and Communication Technology, Electronics and Microelectronics, MIPRO 2014, Opatija, Croatia, May 26–30, 2014, IEEE, 2014, pp. 1300–1303. Available from: https://doi.org/10.1109/MIPRO.2014.6859768.
[2] S.M. Meystre, F.J. Friedlin, B.R. South, S. Shen, M.H. Samore, Automatic de-identification of textual documents in the electronic health record: a review of recent research, BMC Med. Res. Methodol. 10 (1) (2010). Available from: https://doi.org/10.1186/1471-2288-10-70.
[3] A. Stubbs, C. Kotfila, Ö. Uzuner, Automated systems for the de-identification of longitudinal clinical narratives: overview of 2014 i2b2/uthealth shared task track 1, J. Biomed. Inform. 58 (2015) S11–S19. Available from: https://doi.org/10.1016/j.jbi.2015.06.007.

[4] V. Yadav, S. Bethard, A survey on recent advances in named entity recognition from deep learning models, in: E.M. Bender, L. Derczynski, P. Isabelle (Eds.), Proceedings of the 27th International Conference on Computational Linguistics, COLING 2018, Santa Fe, NM, August 20–26, 2018, Association for Computational Linguistics, 2018, pp. 2145–2158. Available from: https://www.aclweb.org/anthology/C18-1182/.

[5] D. Nadeau, S. Sekine, A survey of named entity recognition and classification, Benjamins Current Topics, John Benjamins Publishing Company, 2009, pp. 3–28. Available from: https://doi.org/10.1075/bct.19.03nad.

[6] F. Dernoncourt, J.Y. Lee, Ö. Uzuner, P. Szolovits, De-identification of patient notes with recurrent neural networks, J. Am. Med. Inform. Assoc. 24 (3) (2017) 596–606. Available from: https://doi.org/10.1093/jamia/ocw156.

[7] Z. Huang, W. Xu, K. Yu, Bidirectional LSTM-CRF models for sequence tagging, CoRR (2015)abs/1508.01991. Available from: http://arxiv.org/abs/1508.01991.

[8] J.P.C. Chiu, E. Nichols, Named entity recognition with bidirectional lstm-cnns, Trans. Assoc. Comput. Linguist. 4 (2016) 357–370. Available from: https://transacl.org/ojs/index.php/tacl/article/view/792.

[9] G. Lample, M. Ballesteros, S. Subramanian, K. Kawakami, C. Dyer, Neural architectures for named entity recognition, in: K. Knight, A. Nenkova, O. Rambow (Eds.), NAACL HLT 2016, The 2016 Conference of the North American Chapter of the Association for Computational Linguistics: Human Language Technologies, San Diego, CA, June 12–17, 2016, The Association for Computational Linguistics, 2016, pp. 260–270. Available from: https://doi.org/10.18653/v1/n16-1030.

[10] X. Ma, E.H. Hovy, End-to-end sequence labeling via bi-directional lstm-cnns-crf, Proceedings of the 54th Annual Meeting of the Association for Computational Linguistics, ACL 2016, Berlin, Germany, August 7–12, 2016, vol. 1, Long Papers, The Association for Computer Linguistics, 2016. Available from: https://doi.org/10.18653/v1/p16-1101.

[11] Z. Liu, B. Tang, X. Wang, Q. Chen, De-identification of clinical notes via recurrent neural network and conditional random field, J. Biomed. Inform. 75 (2017) S34–S42. Available from: https://doi.org/10.1016/j.jbi.2017.05.023.

[12] T. Mikolov, I. Sutskever, K. Chen, G.S. Corrado, J. Dean, Distributed representations of words and phrases and their compositionality, in: C.J.C. Burges, L. Bottou, Z. Ghahramani, K.Q. Weinberger (Eds.), Advances in Neural Information Processing Systems 26: 27th Annual Conference on Neural Information Processing Systems 2013. Proceedings of a Meeting Held December 5–8, 2013, Lake Tahoe, NV, 2013, pp. 3111–3119. Available from: http://papers.nips.cc/paper/5021-distributed-representations-of-words-and-phrases-and-their-compositionality.

[13] J.L. Elman, Finding structure in time, Cogn. Sci. 14 (2) (1990) 179–211. Available from: https://doi.org/10.1207/s15516709cog1402_1.

[14] C. Goller, A. Küchler, Learning task-dependent distributed representations by back-propagation through structure, Proceedings of International Conference on Neural Networks (ICNN'96), Washington, DC, June 3–6, 1996, IEEE, 1996, pp. 347–352. Available from: https://doi.org/10.1109/ICNN.1996.548916.

[15] Y. Wu, M. Jiang, J. Lei, H. Xu, Named entity recognition in chinese clinical text using deep neural network, in: I.N. Sarkar, A. Georgiou, P.M. de Azevedo Marques (Eds.), MEDINFO 2015: eHealth-enabled Health — Proceedings of the 15th World Congress on Health and Biomedical Informatics, São Paulo, Brazil, 19–23 August

2015, Studies in Health Technology and Informatics, vol. 216, IOS Press, 2015, pp. 624−628. Available from: https://doi.org/10.3233/978-1-61499-564-7-624.

[16] Y. Wu, J. Xu, M. Jiang, Y. Zhang, H. Xu, A study of neural word embeddings for named entity recognition in clinical text, AMIA 2015, American Medical Informatics Association Annual Symposium, San Francisco, CA, November 14−18, 2015, AMIA, 2015. Available from: http://knowledge.amia.org/59310-amia-1.2741865/t004-1.2745466/f004-1.2745467/2249008-1.2745489/2248738-1.2745486.

[17] Y. Wu, M. Jiang, J. Xu, D. Zhi, H. Xu, Clinical named entity recognition using deep learning models, AMIA 2017, American Medical Informatics Association Annual Symposium, Washington, DC, November 4−8, 2017, AMIA, 2017. Available from: http://knowledge.amia.org/65881-amia-1.3897810/t003-1.3901461/f003-1.3901462/2730946-1.3901506/2731659-1.3901503.

[18] Y. Wu, X. Yang, J. Bian, Y. Guo, H. Xu, W.R. Hogan, Combine factual medical knowledge and distributed word representation to improve clinical named entity recognition, AMIA 2018, American Medical Informatics Association Annual Symposium, San Francisco, CA, November 3−7, 2018, AMIA, 2018. Available from: http://knowledge.amia.org/67852-amia-1.4259402/t004-1.4263758/t004-1.4263759/2977069-1.4263781/2976467-1.4263778.

[19] F. Dernoncourt, J.Y. Lee, P. Szolovits, Neuroner: an easy-to-use program for named-entity recognition based on neural networks, in: L. Specia, M. Post, M. Paul (Eds.), Proceedings of the 2017 Conference on Empirical Methods in Natural Language Processing, EMNLP 2017, Copenhagen, Denmark, September 9−11, 2017, System Demonstrations, Association for Computational Linguistics, 2017, pp. 97−102. Available from: https://doi.org/10.18653/v1/d17-2017.

[20] A. Vaswani, N. Shazeer, N. Parmar, J. Uszkoreit, L. Jones, A.N. Gomez, et al., Attention is all you need, in: I. Guyon, U. von Luxburg, S. Bengio, H.M. Wallach, R. Fergus, S.V.N. Vishwanathan, et al. (Eds.), Advances in Neural Information Processing Systems 30: Annual Conference on Neural Information Processing Systems 2017, Long Beach, CA, December 4−9, 2017, 2017, pp. 5998−6008. Available from: http://papers.nips.cc/paper/7181-attention-is-all-you-need.

[21] J. Devlin, M. Chang, K. Lee, K. Toutanova, BERT: pre-training of deep bidirectional transformers for language understanding, in: J. Burstein, C. Doran, T. Solorio (Eds.), Proceedings of the 2019 Conference of the North American Chapter of the Association for Computational Linguistics: Human Language Technologies, NAACL-HLT 2019, Minneapolis, MN, June 2−7, 2019, vol. 1, Long and Short Papers, Association for Computational Linguistics, 2019, pp. 4171−4186. Available from: https://doi.org/10.18653/v1/n19-1423.

[22] E. Alsentzer, J. Murphy, W. Boag, W.H. Weng, D. Jindi, T. Naumann, et al., Publicly available clinical BERT embeddings, Proceedings of the 2nd Clinical Natural Language Processing Workshop, Minneapolis, MN, Association for Computational Linguistics, 2019, pp. 72−78. Available from: 10.18653/v1/W19-1909. Available from: https://www.aclweb.org/anthology/W19-1909.

[23] J. Lee, W. Yoon, S. Kim, D. Kim, S. Kim, C.H. So, et al., Biobert: a pre-trained biomedical language representation model for biomedical text mining, Bioinform 36 (4) (2020) 1234−1240. Available from: https://doi.org/10.1093/bioinformatics/btz682.

[24] M. Marimon, A. Gonzalez-Agirre, A. Intxaurrondo, H. Rodriguez, J.L. Martin, M. Villegas, et al., Automatic de-identification of medical texts in spanish: the

MEDDOCAN track, corpus, guidelines, methods and evaluation of results, in: M.Á. G. Cumbreras, J. Gonzalo, E.M. Cámara, R. Martnez-Unanue, P. Rosso, J. Carrillo-de-Albornoz, et al. (Eds.), Proceedings of the Iberian Languages Evaluation Forum Co-Located with 35th Conference of the Spanish Society for Natural Language Processing, IberLEF@SEPLN 2019, Bilbao, Spain, September 24, 2019, CEUR Workshop Proceedings, vol. 2421, CEUR-WS.org, 2019, pp. 618–638. Available from: http://ceur-ws.org/Vol-2421/MEDDOCAN_overview.pdf.

[25] B. Tang, D. Jiang, Q. Chen, X. Wang, J. Yan, Y. Shen, De-identification of clinical text via bi-lstm-crf with neural language models, AMIA 2019, American Medical Informatics Association Annual Symposium, Washington, DC, November 16–20, 2019, AMIA, 2019. Available from: http://knowledge.amia.org/69862-amia-1.4570936/t004-1.4574923/t004-1.4574924/3203046-1.4574964/3201562-1.4574961.

[26] J. Pennington, R. Socher, C.D. Manning, Glove: Global vectors for word representation, in: A. Moschitti, B. Pang, W. Daelemans (Eds.), Proceedings of the 2014 Conference on Empirical Methods in Natural Language Processing, EMNLP 2014, Doha, Qatar, October 25–29, 2014, A Meeting of SIGDAT, a Special Interest Group of the ACL, ACL, 2014, pp. 1532–1543. Available from: https://doi.org/10.3115/v1/d14-1162.

[27] P. Bojanowski, E. Grave, A. Joulin, T. Mikolov, Enriching word vectors with subword information, Trans. Assoc. Comput. Linguist. 5 (2017) 135–146. Available from: https://transacl.org/ojs/index.php/tacl/article/view/999.

[28] M.E. Peters, M. Neumann, M. Iyyer, M. Gardner, C. Clark, K. Lee, et al., Deep contextualized word representations, in: M.A. Walker, H. Ji, A. Stent (Eds.), Proceedings of the 2018 Conference of the North American Chapter of the Association for Computational Linguistics: Human Language Technologies, NAACL-HLT 2018, New Orleans, LA, June 1–6, 2018, vol. 1, Long Papers, Association for Computational Linguistics, 2018, pp. 2227–2237. Available from: https://doi.org/10.18653/v1/n18-1202.

[29] A. Akbik, D. Blythe, R. Vollgraf, Contextual string embeddings for sequence labeling, in: E.M. Bender, L. Derczynski, P. Isabelle (Eds.), Proceedings of the 27th International Conference on Computational Linguistics, COLING 2018, Santa Fe, NM, August 20–26, 2018, Association for Computational Linguistics, 2018, pp. 1638–1649. Available from: https://www.aclweb.org/anthology/C18-1139/.

[30] A. Radford, J. Wu, R. Child, D. Luan, D. Amodei, I. Sutskever, Language models are unsupervised multitask learners, OpenAI Blog (2019).

[31] M. Schuster, K. Nakajima, Japanese and korean voice search, 2012 IEEE International Conference on Acoustics, Speech and Signal Processing, ICASSP 2012, Kyoto, Japan, March 25–30, 2012, IEEE, 2012, pp. 5149–5152. Available from: https://doi.org/10.1109/ICASSP.2012.6289079.

[32] M.E. Peters, M. Neumann, L. Zettlemoyer, W. Yih, Dissecting contextual word embeddings: architecture and representation, in: E. Riloff, D. Chiang, J. Hockenmaier, J. Tsujii (Eds.), Proceedings of the 2018 Conference on Empirical Methods in Natural Language Processing, Brussels, Belgium, October 31–November 4, 2018, Association for Computational Linguistics, 2018, pp. 1499–1509. Available from: https://doi.org/10.18653/v1/d18-1179.

[33] U. Khandelwal, H. He, P. Qi, D. Jurafsky, Sharp nearby, fuzzy far away: how neural language models use context, in: I. Gurevych, Y. Miyao (Eds.), Proceedings of the

56th Annual Meeting of the Association for Computational Linguistics, ACL 2018, Melbourne, Australia, July 15−20, 2018, vol. 1, Long Papers, Association for Computational Linguistics, 2018, pp. 284−294. Available from: https://www.aclweb.org/anthology/P18-1027/.

[34] K. Pantazos, S. Lauesen, S. Lippert, Preserving medical correctness, readability and consistency in de-identified health records, Health Inform. J. 23 (4) (2017) 291−303. Available from: https://doi.org/10.1177/1460458216647760.

[35] E. Scheurwegs, K. Luyckx, F.V. der Schueren, T.V. den Bulcke, De-identification of clinical free text in dutch with limited training data: a case study, in: G. Savova, K. B. Cohen, G. Angelova (Eds.), Proceedings of the Workshop on NLP for Medicine and Biology Associated with RANLP 2013, Hissar, Bulgaria, September 13, 2013, INCOMA Ltd., Shoumen, Bulgaria, 2013, pp. 18−23. Available from: https://www.aclweb.org/anthology/W13-5103/.

[36] J. Trienes, D. Trieschnigg, C. Seifert, D. Hiemstra, Comparing rule-based, feature-based and deep neural methods for de-identification of dutch medical records, in: C. Eickhoff, Y. Kim, R.W. White (Eds.), Proceedings of the ACM WSDM 2020 Health Search and Data Mining Workshop, Co-Located with the 13th ACM International WSDM Conference, HSDM@WSDM 2020, Houston, TX, February 3, 2020, CEUR Workshop Proceedings, vol. 2551, CEUR-WS.org, 2020, pp. 3−11. Available from: http://ceur-ws.org/Vol-2551/paper-03.pdf.

[37] C. Grouin, A. Névéol, De-identification of clinical notes in french: towards a protocol for reference corpus development, J. Biomed. Inform. 50 (2014) 151−161. Available from: https://doi.org/10.1016/j.jbi.2013.12.014.

[38] C. Gaudet-Blavignac, V. Foufi, E. Wehrli, C. Lovis, De-identification of french medical narratives, Swiss Med. Inform. (2018). Available from: https://doi.org/10.4414/smi.34.00417.

[39] K. Tomanek, P. Daumke, F. Enders, J. Huber, K. Theres, M. Müller, An interactive de-identification-system, Proc. SMBM (2012) 82−86.

[40] P. Richter-Pechanski, S. Riezler, C. Dieterich, De-identification of german medical admission notes, in: U. Hübner, U. Sax, H. Prokosch, B. Breil, H. Binder, A. Zapf, et al. (Eds.), German Medical Data Sciences: A Learning Healthcare System − Proceedings of the 63rd Annual Meeting of the German Association of Medical Informatics, Biometry and Epidemiology (GMDS e.V.) 2018 in Osnabrück, Germany, September 2−6, 2018, GMDS 2018, Studies in Health Technology and Informatics, vol. 253, IOS Press, 2018, pp. 165−169. Available from: https://doi.org/10.3233/978-1-61499-896-9-165.

[41] A. Tveit, O. Edsberg, T. Rost, A. Faxvaag, O. Nytro, T. Nordgard, et al., Anonymization of general practitioner medical records, Second HelsIT Conference, 2004.

[42] M. Marciniak, A. Mykowiecka, P. Rychlik, Medical text data anonymization, J. Med. Inform. Technol. 16 (2010).

[43] P. Borowik, P. Brylicki, M. Dzieciatko, W. Jeda, L. Leszewski, P. Zajac, De-identification of electronic health records data, in: E. Pietka, P. Badura, J. Kawa, W. Wieclawek (Eds.), Information Technology in Biomedicine, ITIB 2019, Kamień Ślaski, Poland, June 18−20, 2019, Advances in Intelligent Systems and Computing, vol. 1011, Springer, 2019, pp. 325−337. Available from: https://doi.org/10.1007/978-3-030-23762-2_29.

[44] N.J. Mamede, J. Baptista, F. Dias, Automated anonymization of text documents, IEEE Congress on Evolutionary Computation, CEC 2016, Vancouver, BC, Canada, July 24−29, 2016, IEEE, 2016, pp. 1287−1294. Available from: https://doi.org/10.1109/CEC.2016.7743936.

[45] A. Alfalahi, S. Brissman, H. Dalianis, Pseudonymisation of personal names and other phis in an annotated clinical swedish corpus, LREC 2012, Istanbul, Turkey, May 23−25, 2012.

[46] S.J. Pan, Q. Yang, A survey on transfer learning, IEEE Trans. Knowl. Data Eng. 22 (10) (2010) 1345−1359. Available from: https://doi.org/10.1109/TKDE.2009.191.

[47] J. Kim, Y. Kim, R. Sarikaya, E. Fosler-Lussier, Cross-lingual transfer learning for POS tagging without cross-lingual resources, in: M. Palmer, R. Hwa, S. Riedel (Eds.), Proceedings of the 2017 Conference on Empirical Methods in Natural Language Processing, EMNLP 2017, Copenhagen, Denmark, September 9−11, 2017, Association for Computational Linguistics, 2017, pp. 2832−2838. Available from: https://doi.org/10.18653/v1/d17-1302.

[48] J. Xie, Z. Yang, G. Neubig, N.A. Smith, J.G. Carbonell, Neural cross-lingual named entity recognition with minimal resources, in: E. Riloff, D. Chiang, J. Hockenmaier, J. Tsujii (Eds.), Proceedings of the 2018 Conference on Empirical Methods in Natural Language Processing, Brussels, Belgium, October 31−November 4, 2018, Association for Computational Linguistics, 2018, pp. 369−379. Available from: https://doi.org/10.18653/v1/d18-1034.

[49] W.U. Ahmad, Z. Zhang, X. Ma, E.H. Hovy, K. Chang, N. Peng, On difficulties of cross-lingual transfer with order differences: a case study on dependency parsing, in: J. Burstein, C. Doran, T. Solorio (Eds.), Proceedings of the 2019 Conference of the North American Chapter of the Association for Computational Linguistics: Human Language Technologies, NAACL-HLT 2019, Minneapolis, MN, June 2−7, 2019, vol. 1, Long and Short Papers, Association for Computational Linguistics, 2019, pp. 2440−2452. Available from: https://doi.org/10.18653/v1/n19-1253.

[50] B. Heinzerling, M. Strube, Bpemb: Tokenization-free pre-trained subword embeddings in 275 languages, in: N. Calzolari, K. Choukri, C. Cieri, T. Declerck, S. Goggi, K. Hasida, et al. (Eds.), Proceedings of the Eleventh International Conference on Language Resources and Evaluation, LREC 2018, Miyazaki, Japan, May 7−12, 2018, European Language Resources Association (ELRA), 2018. Available from: http://www.lrec-conf.org/proceedings/lrec2018/summaries/1049.html.

[51] R. Sennrich, B. Haddow, A. Birch, Neural machine translation of rare words with subword units, Proceedings of the 54th Annual Meeting of the Association for Computational Linguistics, ACL 2016, Berlin, Germany, August 7−12, 2016, vol. 1, Long Papers, The Association for Computer Linguistics, 2016. Available from: https://doi.org/10.18653/v1/p16-1162.

[52] J. Bingel, J. Bjerva, Cross-lingual complex word identification with multitask learning, in: J.R. Tetreault, J. Burstein, E. Kochmar, C. Leacock, H. Yannakoudakis (Eds.), Proceedings of the Thirteenth Workshop on Innovative Use of NLP for Building Educational Applications@NAACL-HLT 2018, New Orleans, LA, June 5, 2018, Association for Computational Linguistics, 2018, pp. 166−174. Available from: https://doi.org/10.18653/v1/w18-0518.

[53] S.M. Yimam, C. Biemann, S. Malmasi, G. Paetzold, L. Specia, S. Stajner, et al., A report on the complex word identification shared task 2018, in: J.R. Tetreault, J.

Burstein, E. Kochmar, C. Leacock, H. Yannakoudakis (Eds.), Proceedings of the Thirteenth Workshop on Innovative Use of NLP for Building Educational Applications@NAACL-HLT 2018, New Orleans, LA, June 5, 2018, Association for Computational Linguistics, 2018, pp. 66–78. Available from: https://doi.org/10.18653/v1/w18-0507.

[54] M. Zhao, H. Schütze, A multilingual BPE embedding space for universal sentiment lexicon induction, in: A. Korhonen, D.R. Traum, L. Màrquez (Eds.), Proceedings of the 57th Conference of the Association for Computational Linguistics, ACL 2019, Florence, Italy, July 28–August 2, 2019, vol. 1, Long Papers, Association for Computational Linguistics, 2019, pp. 3506–3517. Available from: https://doi.org/10.18653/v1/p19-1341.

[55] A. Akbik, T. Bergmann, D. Blythe, K. Rasul, S. Schweter, R. Vollgraf, FLAIR: an easy-to-use framework for state-of-the-art NLP, in: W. Ammar, A. Louis, N. Mostafazadeh (Eds.), Proceedings of the 2019 Conference of the North American Chapter of the Association for Computational Linguistics: Human Language Technologies, NAACL-HLT 2019, Minneapolis, MN, June 2–7, 2019, Demonstrations, Association for Computational Linguistics, 2019, pp. 54–59. Available from: https://doi.org/10.18653/v1/n19-4010.

[56] P. Mulcaire, J. Kasai, N.A. Smith, Polyglot contextual representations improve cross-lingual transfer, in: J. Burstein, C. Doran, T. Solorio (Eds.), Proceedings of the 2019 Conference of the North American Chapter of the Association for Computational Linguistics: Human Language Technologies, NAACL-HLT 2019, Minneapolis, MN, June 2–7, 2019, vol. 1, Long and Short Papers, Association for Computational Linguistics, 2019, pp. 3912–3918. Available from: https://doi.org/10.18653/v1/n19-1392.

[57] M. Arkhipov, M. Trofimova, Y. Kuratov, A. Sorokin, Tuning multilingual transformers for language-specific named entity recognition, Proceedings of the 7th Workshop on Balto-Slavic Natural Language Processing, Association for Computational Linguistics, 2019. Available from: https://doi.org/10.18653/v1/w19-3712.

[58] W.L. Taylor, "Cloze procedure": a new tool for measuring readability, Journal. Mass Commun. Q. 30 (4) (1953) 415–433. Available from: https://doi.org/10.1177/107769905303000401.

[59] T. Pires, E. Schlinger, D. Garrette, How multilingual is multilingual bert? in: A. Korhonen, D.R. Traum, L. Màrquez (Eds.), Proceedings of the 57th Conference of the Association for Computational Linguistics, ACL 2019, Florence, Italy, July 28–August 2, 2019, vol. 1, Long Papers, Association for Computational Linguistics, 2019, pp. 4996–5001. Available from: https://doi.org/10.18653/v1/p19-1493.

[60] S. Wu, M. Dredze, Beto, bentz, becas: the surprising cross-lingual effectiveness of BERT, in: K. Inui, J. Jiang, V. Ng, X. Wan (Eds.), Proceedings of the 2019 Conference on Empirical Methods in Natural Language Processing and the 9th International Joint Conference on Natural Language Processing, EMNLP-IJCNLP 2019, Hong Kong, China, November 3–7, 2019, Association for Computational Linguistics, 2019, pp. 833–844. Available from: https://doi.org/10.18653/v1/D19-1077.

[61] K. Karthikeyan, Z. Wang, S. Mayhew, D. Roth, Cross-lingual ability of multilingual BERT: an empirical study, 8th International Conference on Learning Representations, ICLR 2020, Addis Ababa, Ethiopia, April 26–30, 2020, OpenReview.net, 2020. Available from: https://openreview.net/forum?id = HJeT3yrtDr.

[62] B. Heinzerling, M. Strube, Sequence tagging with contextual and non-contextual sub-word representations: a multilingual evaluation, in: A. Korhonen, D.R. Traum, L. Màrquez (Eds.), Proceedings of the 57th Conference of the Association for Computational Linguistics, ACL 2019, Florence, Italy, July 28–August 2, 2019, vol. 1, Long Papers, Association for Computational Linguistics, 2019, pp. 273–291. Available from: https://doi.org/10.18653/v1/p19-1027.

[63] L.A. Ramshaw, M. Marcus, Text chunking using transformation-based learning, in: D. Yarowsky, K. Church (Eds.), Third Workshop on Very Large Corpora, VLC@ACL 1995, Cambridge, MA, June 30, 1995, 1995. Available from: https://www.aclweb.org/anthology/W95-0107/.

[64] A. Stubbs, Ö. Uzuner, Annotating longitudinal clinical narratives for de-identification: the 2014 i2b2/uthealth corpus, J. Biomed. Inform. 58 (2015) S20–S29. Available from: https://doi.org/10.1016/j.jbi.2015.07.020.

[65] R. Catelli, F. Gargiulo, V. Casola, G.D. Pietro, H. Fujita, M. Esposito, A novel COVID-19 data set and an effective deep learning approach for the de-identification of italian medical records, IEEE Access. 9 (2021) 19097–19110. Available from: https://doi.org/10.1109/ACCESS.2021.3054479.

[66] R. Catelli, F. Gargiulo, V. Casola, G.D. Pietro, H. Fujita, M. Esposito, Crosslingual named entity recognition for clinical de-identification applied to a COVID-19 italian data set, Appl. Soft Comput. 97 (Part) (2020) 106779. Available from: https://doi.org/10.1016/j.asoc.2020.106779.

[67] R. Catelli, F. Gargiulo, E. Damiano, M. Esposito, G.D. Pietro, Clinical de-identification using sub-document analysis and ELECTRA, IEEE International Conference on Digital Health, ICDH 2021, Chicago, IL, September 5–10, 2021, IEEE, 2021, pp. 266–275. Available from: https://doi.org/10.1109/ICDH52753.2021.00050.

[68] R. Catelli, L. Bevilacqua, N. Mariniello, V. Scotto di Carlo, M. Magaldi, H. Fujita, et al., Cross lingual transfer learning for sentiment analysis of italian tripadvisor reviews, Expert. Syst. Appl. 209 (2022) 118246. Available from: https://doi.org/10.1016/j.eswa.2022.118246, https://www.sciencedirect.com/science/article/pii/S0957417422013926.

[69] R. Catelli, H. Fujita, G. De Pietro, M. Esposito, Deceptive reviews and sentiment polarity: effective link by exploiting bert, Expert. Syst. Appl. 209 (2022) 118290. Available from: https://doi.org/10.1016/j.eswa.2022.118290, https://www.sciencedirect.com/science/article/pii/S0957417422014269.

Estimation of COVID-19 fatality associated with different SARS-CoV-2 variants

8

Marco Pota[1], Andrea Pota[2] and Massimo Esposito[3]

[1]*Institute for High Performance Computing and Networking (ICAR)—National Research Council of Italy (CNR), Napoli, Italy*
[2]*UOC Nefrologia e Dialisi Ospedale del Mare, Napoli, Italy*
[3]*Institute for High Performance Computing and Networking (ICAR), National Research Council (CNR), Naples, Italy*

8.1 Introduction

Last year, the research community addressed most of its efforts to the pandemic issues, since social, economic, and healthcare settings of many countries were hard tested by the new coronavirus. SARS-CoV-2 is among the most infective viruses known, for human beings, indeed it acquired the characteristics of a pandemic, and as of the end of March 2022, it has infected more than 486 million people all over the world [1]. Consequently, SARS-CoV-2 is dangerous for its large case count, although less fatal with respect to other coronaviruses such as SARS-CoV and MERS-CoV [2]. On the other hand, the clinical expression of this virus, named COVID-19 or generically coronavirus disease, is important for public health, according to the heavy damage for patients and relatively high probability of death. As a consequence of both high infectivity and high fatality, the number of deaths associated with COVID-19 reported as of the end of March 2022 is more than 6 million, increasing every day, as reported in data released online by World Health Organization (WHO) [1].

As time goes by, the knowledge about COVID-19 disease is increasing. Prevention policies have been adopted in most countries (e.g., Refs. [3,4]). Moreover, different types of tests are available to the majority of the people, and policies are adopted by many governments to incentivize their use, particularly for cases at risk. Treatments for clinical care and management of serious cases have been developed (e.g., Refs. [5–18]), and some are well established [19,20]. Furthermore, 14 types of vaccines have been released, as detailed in the following.

Nevertheless, despite the great efforts of scientific, medical, pharmaceutical, and political communities, for prevention, monitoring, management, and care, the

Artificial Intelligence in Healthcare and COVID-19. DOI: https://doi.org/10.1016/B978-0-323-90531-2.00013-8

infection is still spreading, with hundreds of thousands of new cases on a daily basis [1], and a high number of deaths still occur [1]. In particular, the emphasis of the global healthcare emergency depends on two main factors: (1) the high transmissibility of the SARS-CoV-2 virus, which implies a high number of people infected, and (2) the high number of deaths among infected people.

On the other hand, the virus evolved over time, and some variants have been defined as variants of concern (VOC) [21]. In particular, both the transmissibility of the virus and the death rate of the disease can change between one variant and another.

In this study, the epidemiological side of the issue is taken into account. In particular, it is focused on mortality.

Different studies already calculated estimates of death rate associated with this virus, but most of them have drawbacks, as detailed in the next subsection. Moreover, many works aimed at a single estimation, at a single time instant, or averaged over time.

Instead, this work presents the evolution of mortality from the beginning of the pandemic until nowadays, in order to individuate eventual peculiar behaviors. Moreover, the evolution of the mortality is correlated with the changes in the predominant variants, in order to associate peculiar behaviors with different VOC. This analysis could indicate a tightening or a softening of the mortality, could enhance the knowledge about the problem, and suggest the most appropriate strategies needed to manage the emergency.

To this aim, available data and previously proposed methods to estimate mortality have been employed.

Data concerning the Italian situation in terms of infections and deaths have been chosen, which the Civil Protection Department of the Presidency of the Council of Ministers of Italy is making available on a daily basis [22]. At the same time, data regarding the prevalence of different variants in Italy have been taken from communications of the Istituto Superiore di Sanità, summarizing the responses of sequencing of patients' samples [23]. In the following, after the data description, possible uncertainties (particularly due to underestimation of positive cases) of the reports are discussed, and the hypotheses needed to generalize conclusions obtained from these data on different situations are stated.

As long as the statistical methods to estimate mortality are concerned, two main features could be of interest: the case fatality ratio (CFR, number of deaths due to COVID-19 with respect to the number of COVID-19 patients) and infection fatality ratio (IFR, number of deaths due to COVID-19 with respect to the number of SARS-CoV-2 infected people). IFR is considered here instead of CFR, since all the available data report positive cases as those including all confirmed SARS-CoV-2 infected people, without distinguishing between asymptomatic cases and COVID-19 patients with subclinical or more serious symptoms. Different methods to calculate IFR are discussed in the following, and all of them have been used here to evaluate the time evolution of fatality.

The rest of the paper is structured as follows: the following subsection outlines related work, and Section 8.2 reports details about the data and the methods used to estimate IFR. Then, Section 8.3 presents the obtained results, and Section 8.4 discusses them and draws conclusions.

8.1.1 Related work

Different studies reported estimates of COVID-19 mortality in terms of CFR or IFR.

Some of them obtained mortality by dividing confirmed deaths by confirmed infected people. An early study reported 6 deaths over a limited cohort of 41 confirmed positive cases (CFR equal to 14.6%), and an overall IFR of about 3% in a wider sample in Wuhan [24]. However, the authors stated: "both of these estimates should be treated with great caution because not all patients have concluded their illness (i.e., recovered or died) and the true number of infections and full disease spectrum are unknown." Moreover, WHO reported, on January 29, 2020, "2% of confirmed cases are reported to have died" [25]. Later, on February 28, 2020, the WHO–China Joint Mission reported [26] that 2114 people died over 55,924 studied confirmed cases, which corresponds to an IFR of 3.8%. However, "The Joint Mission acknowledges the known challenges and biases of reporting crude CFR early in an epidemic." The same report underlined dependence on risk factors and location. Finally, on March 3, 2020, the WHO Director-General stated: "Globally, about 3.4% of reported COVID-19 cases have died" [27]. Moreover, another work reported a CFR of 7.05%, calculated by confirmed cases updated on April 30, 2020 [28]; this was actually an IFR estimate since employed data included in confirmed cases all SARS-CoV-2 infected people, both COVID-19 patients and asymptomatic.

Other studies adjusted the calculation by only considering resolved cases, and calculating IFR as the number of deaths divided by the sum of deaths and recovered people, as suggested in Ref. [29]. For example, in Ref. [30], an IFR of 1.4% was found based on New York City adjusted data mentioned above, while values of 4% worldwide and 3.5% outside mainland China were found based on confirmed cases only [31].

An alternative method to calculate IFR, during an ongoing epidemic, is to consider daily cases, as suggested in Ref. [29]. In particular, in Ref. [32], statistical methods were employed to derive a time-delayed adjusted IFR estimate lower than 0.5%.

One of the latest works [33] estimated infections based on seroprevalence, although it is biased by the fact that people doing tests have a higher probability of being infected, and used a meta-regression model to calculate IFR, identifying differences by age, time, and geography.

While all the previously mentioned methods imply assumptions about the estimation of infected people, in Ref. [34], an alternative method was proposed, based on daily cases, which takes into account only days with high number of tests performed per death, in order to obtain the best estimate of infected people. The same work calculated an IFR of about 0.9%.

All these past studies obtained definitive estimations of IFR. On the contrary, IFR variations in time are often neglected, and very few works reported differences among variants, some founding no overall difference between the fatality of the Delta variant and the original virus [35], others comparing vaccine effectiveness between Omicron and Delta variants [36], others founding Omicron variant less severe with respect to Delta [36].

In this work, four different statistical methods are employed to obtain the temporal evolution of IFR and individuate interesting trends, particularly correlated with all the predominant variants, and to obtain an updated estimate for comparison with the older ones.

8.2 Materials and methods

In this section, first data employed for this study are detailed, regarding infections and deaths (2.1), and regarding predominant variants (2.2) in Italy. Moreover, the methods used here to calculate IFR are reported (2.3). Finally, preliminary observations about uncertainties in data and IFR calculation (2.4), about dependence on vaccine distribution (2.5), and about hypotheses needed to generalize findings (2.6) are reported.

8.2.1 Data on COVID-19 infections and deaths

The data set used here regards the COVID-19 situation in Italy and is released on GitHub [22] by the Civil Protection Department of the Presidency of the Council of Ministers of Italy. Data start from the recognized beginning of the infections, that is, February 24, 2020, and are updated on a daily basis. Both the data about the entire nation and each region and province are included. However, for the aims of this work, only data about the whole Italy are considered.

Numerous features are reported, regarding positive cases, recovered cases, deaths, and tests performed. In this work, the following are used: date, new positives (new SARS-CoV-2 positives at date), cumulative recovered (total SARS-CoV-2 former positives currently negative), cumulative deaths (total deaths due to COVID-19), and cumulative cases (total number of confirmed SARS-CoV-2 positive cases). Moreover, the further features of new deaths (new deaths due to COVID-19 till date) and new tests (new tests performed on date) are simply calculated by daily increments of cumulative deaths and cumulative tests, respectively. In Table 8.1, a sample of data used here is reported, in correspondence with some notable dates (the first day reported, the day with the highest number of new deaths, the one with the highest number of new positives and new tests, and the most recent date available).

The presented features enable to draw the time series of IFR, calculated for each date as detailed below.

Table 8.1 Examples of daily data used.

Date	New positives	Cumulative cases	Cumulative recovered	New deaths	Cumulative deaths	New tests
February 24, 2020	221	229	1	7	7	4324
December 3, 2020	23,225	1,664,829	846,809	993	58,038	226,729
January 18, 2022	212,004	9,018,425	6,314,444	434	141,825	1,481,349
March 29, 2022	99,457	14,496,579	13,070,647	177	159,054	660,708

8.2.2 Data about SARS-CoV-2 variants

SARS-CoV-2 virus, like all the others, is subject to mutations. While most changes with respect to the original virus have little impact on its behavior, some others imply changes in contagiousness, severity of the disease, effectiveness of vaccines and therapies, and/or performance of diagnostic tools. In particular, IFR could be affected by virus mutations. Therefore, WHO encourages authorities to keep surveillance by performing virus sequencing to provide indications of the prevalence of SARS-CoV-2 variants and to detect new ones.

Given the high number of possible mutations, the WHO Technical Advisory Group on Virus Evolution recommended using letters of the Greek alphabet to name the most important variants. Moreover, some variants have been classified as variants of interest (VOI), and some of them as VOC. A VOI is defined [37] as "A SARS-CoV-2 variant: (i) with genetic changes that are predicted or known to affect virus characteristics such as transmissibility, disease severity, immune escape, diagnostic or therapeutic escape; and (ii) identified to cause significant community transmission or multiple COVID-19 clusters, in multiple countries with increasing relative prevalence alongside increasing number of cases over time, or other apparent epidemiological impacts to suggest an emerging risk to global public health." The VOI identified up to now are the following: Alpha, Beta, Gamma, Delta, Epsilon, Zeta, Eta, Theta, Iota, Kappa, Lambda, Mu, and Omicron [37]. A VOC is defined [37] as "A SARS-CoV-2 variant that meets the definition of a VOI and, through a comparative assessment, has been demonstrated to be associated with one or more of the following changes at a degree of global public health significance: (i) increase in transmissibility or detrimental change in COVID-19 epidemiology; or (ii) increase in virulence or change in clinical disease presentation; or (iii) decrease in the effectiveness of public health and social measures or available diagnostics, vaccines, therapeutics." The following VOI were also included as VOC: Alpha, Beta, Gamma, Delta, and Omicron.

Data about the prevalence of different variants in Italy have not been made available. However, information extracted from communications of the Istituto Superiore di Sanità [23] allows to infer approximate dates when a certain variant became the predominant one in Italy. In Table 8.2, the intervals of dates when

Table 8.2 Intervals of dates when different virus variants have been predominant.

From	To	Variant
February 24, 2020	December 31, 2020	Original
January 1, 2021	June 4, 2021	Alpha
June 5, 2021	January 2, 2022	Delta
January 3, 2022	March 29, 2022	Omicron

each variant has been the most prevalent in Italy are reported, and these intervals are used for this study.

Of course, the infection rate of different variants is expected to increase with time, since generally a variant becomes predominant when it is the one with the highest transmissibility. On the other hand, the fatality of different variants can be evaluated, by correlating these time intervals with the IFR time series.

8.2.3 Models to estimate fatality

As mentioned in the Introduction, different methods exist to estimate IFR.

The first method simply divides the number of deaths due to COVID-19 by the total number of confirmed SARS-CoV-2 positive cases [29]:

$$\text{IFR}_1 = \frac{\text{cumulative deaths}}{\text{cumulative cases}}. \tag{8.1}$$

However, this method gives a wrong estimation, because cumulative cases include unresolved cases [29] and because actual cumulative cases are surely underestimated by confirmed cumulative cases.

A second method allows for mitigating the first problem, by only taking into account resolved cases (comprising deaths due to COVID-19 and recovered cases). It compares the number of deaths with the number of resolved cases [29]:

$$\text{IFR}_2 = \frac{\text{cumulative deaths}}{(\text{cumulative deaths} + \text{cumulative recovered})}. \tag{8.2}$$

However, also in this case, the estimation of actual cumulative recovered is not trivial.

A third method to calculate IFR, suggested in Ref. [29], calculates IFR_1 on a daily basis: similar to IFR_1, where cumulative deaths at the end of the infection are a fraction of cumulative cases, during the infection, the new deaths each day should be the same fraction of the new positives of some days before:

$$\text{IFR}_3 = \frac{\text{new deaths}(i + T)}{\text{new positives}(i)}, \tag{8.3}$$

where i is the variable date, and T is the medium number of days from infection to fatal resolution.

With this method, a definition of T is needed. Luckily, two studies on the development of COVID-19 have shown that, for nonsurvivor patients, death occurs on average during the 19th day after the first symptoms and beginning of inspections [38], while for survivors, the average time from onset of symptoms to negative testing is 19 days [39,40]. Moreover, another study estimated the median period from virus exposure to the first symptoms onset in between 5 and 6 days [40]. Therefore, here, the following T, obtained by summing up the two periods between exposure and death, is chosen:

$$T = 24. \tag{8.4}$$

Actually, the period between infection and eventual death is variable; however, here, this time delay is fixed at its median value, according to literature findings.

On the other hand, this method also needs to estimate actual positive cases. However, while other methods involve cumulative cases, this one uses new positive cases, which can approximate the actual situation.

A further approach [34] proposed to employ the obvious observation that the estimation of positive cases is better the more tests is performed. In particular, IFR is estimated according to Eq. (8.3):

$$IFR_4 = \lim_{\frac{\text{new tests}(i)}{\text{new deaths}(i+T)} \to +\infty} \frac{\text{new deaths}(i+T)}{\text{new positives}(i)}. \tag{8.5}$$

The measure that should tend to infinity was chosen for these reasons: (1) obviously, the estimate of new positives on day i is better the more tests are performed on day i; (2) the quantity of tests should not be evaluated as an absolute value, but with respect to the number of actual positive cases; (3) the number of actual positive cases on day i is unknown, but according to Eq. (8.3) it is proportional to new deaths on day $i + T$. The limit can be evaluated on the basis of the available data. The IFR values approaching the asymptote can be chosen as those corresponding to the highest tests/deaths ratio. In particular, the mean of the last three values is considered, to average errors due to stochastic deviations.

This approach overcomes the estimation of positive cases. However, it allows calculating a single value of IFR, within a certain period, while it is not possible to look at the whole time evolution.

8.2.4 Uncertainty of available data and fatality estimation

The calculation of IFR needs a few numbers for a given population, that is, the number of deaths and the number of infected people, while the number of recovered people is easily obtained as a difference. Although these numbers are reported in data used here, they need some discussion.

The actual number of deaths due to COVID-19 is different from the number of deaths with that cause confirmed, due to some sources of errors. First, the number of confirmed deaths is overestimated, since it comprises people who died after performing a laboratory test, but also people labeled as "probable," without a test confirming their positivity. On the other hand, the number is underestimated, since some positive cases could have died without being hospitalized and labeled as confirmed COVID-19 cases. For example, in New York City, until May 1, 2020, 18,282 deaths were reported, of which 13,156 were confirmed positive cases, and other 5126 were probable cases [41]. On the other hand, a study calculated, on the same date, an excess of 24,172 deaths with respect to the seasonal expected baseline [42].

The actual number of infected people is much more uncertain, and it is certainly underestimated by the number of confirmed cases [43], because the testing has not been applied to the entire population, and there is a large number of asymptomatic cases or people with symptoms mild enough to not require testing [5]. For example, a study performing an antibody test on 15,000 healthy people showed [44] that in New York City, on May 1, 2020, 19.9% of the population had COVID-19 antibodies, which, multiplied by the entire city population (8,398,748 [45]), gives an estimate of 1,671,351 recovered cases, about 10 times the number of confirmed cases at the same date (166,883 [46]).

As a consequence, the result of IFR calculation and its time evolution depends on the following estimation errors.

First, during a period when the infection is still ongoing, the calculation of IFR_1 using current total numbers is an underestimation, due to the unknown outcome of people not yet resolved. However, if the number of current not resolved infections is negligible with respect to cumulative cases, this type of error is negligible. Moreover, some methods overcome this problem by considering only resolved cases (IFR_2) or considering daily cases (IFR_3) [29].

Second, due to underestimation of positive cases, IFR_1, IFR_2, and IFR_3 are overestimated. A method was proposed to overcome this issue (IFR_4) [34], but this allows only calculating overall estimates of IFR. However, in this work, the evolution of IFR over time is expected; thus, the overestimation of IFR is acknowledged, as long as a definitive estimate of IFR is not expected.

The amount of the latter type of error is variable over time, in particular for IFR_3, calculated on a daily basis. Indeed, if the amount of underestimation of positive cases is constant, then IFR overestimation is constant, and the estimates are proportional to the actual IFR. However, generally the underestimation of positive cases, with the consequent amount of IFR overestimation, depends on "waves" of the infection.

Moreover, in particular for IFR_3, the time evolution includes a stochastic variance due to a high number of unpredictable factors. This effect can be mitigated by averaging the data of the last few days.

Once these errors are taken into account, the IFR evolution over time can be correlated to factors influencing it. Different factors interact, which are changes in the population, the treatments, the vaccine distribution, and the succession of variants. In particular, changes in population and treatments are expected just during the first phase of the pandemic, while vaccine distribution and diversity of variants affect the whole time evolution.

8.2.5 Correlation with vaccine distribution

Up to March 2022, 14 vaccines have been released, after assessment finalization, by different pharmaceutical companies, some based on mRNA, for example, mRNABNT162b2 (Comirnaty) produced by BioNTech/Fosun Pharma/Pfizer and mRNA-1273 (Spikevax) by Moderna/NIAID, some based on viral vector, for example, AZD1222—ChAdOx1 (Vaxzevria, ex AstraZeneca) by AstraZeneca/Oxford University and JNJ-78436735—Ad26.COV2.S (Johnson & Johnson) by Janssen Pharmaceutical Companies, others based on protein subunit, for example, NVX-CoV2373 (Nuvaxovid) by Novavax, while others are being developed or their evaluation is still ongoing [47]. More than 11 billion doses have been administered globally [1]. In Italy, vaccine doses have been distributed starting from the end of December 2020.

Consequently, IFR is affected by vaccine distribution, hopefully with a decreasing trend. However, demonstrating the significance of IFR decrease depending on the amount of vaccine doses distributed goes beyond the scope of this work, which focuses on correlation with virus variants. Therefore, here eventual dependence on vaccines is observed just in terms of differences between the periods before and after the first vaccine release, while a detailed discussion about the quantitative correlation is foreseen as future work.

8.2.6 Hypotheses to generalize conclusions

The possibility of generalizing outcomes based on an Italian data set to different situations is conditioned by the satisfaction of the following few hypotheses. Some of them may be reasonable in general, while others depend on the analyzed situation. First, a previous retrospective cohort study [48] allowed authors to conclude that the Black race was not associated with higher in-hospital mortality than the White race. This finding suggests as feasible the hypothesis that in general the prognosis does not depend on the race of the patient, which is equivalent to hypothesize that the death rate does not directly depend on the prevalent race of the considered population. On the other hand, the disparities among population groups, regarding the limited accessibility to public healthcare system for ethnic minorities [49], could be more or less exacerbated with respect to Italy, depending on the socioeconomic and cultural situation. Second, the hypothesis is needed that risk factors (e.g., those found in Refs. [5,6,48,50—56] and reviewed in Ref. [57]) are distributed in the considered population similarly to Italy. This is reasonable as long as situations are analyzed where alimentary and lifestyle habits are comparable to Italy. Moreover, the generalization could be considered valid only for those situations where patients are treated similarly to Italy, thanks to similar clinical guidelines, availability of drugs, and room for hospitalization. Finally, further hypotheses are needed, like similarity with Italy in terms of climatic conditions, pollution, and maybe others based on still unknown influencing factors.

On the other hand, while the above hypotheses are necessary to generalize the IFR values and trends to different situations, their correlation with different variants does not directly depend on the situation, and hypotheses can be relaxed to generalize it.

8.3 Results

In this section, data described in Section 8.2.1 and sampled in Table 8.1 are used, with limitations reported in Section 8.2.4. IFR is calculated according to different methods reported in Section 8.2.3. The following figures report both IFR, as it varies during the ongoing infection, and the time intervals when different variants have been predominant in the same population, reported in Table 8.2 of Section 8.2.2. The trends that can be individuated are generalizable under the hypotheses stated in Section 8.2.6.

Fig. 8.1 reports IFR_1, calculated according to Eq. (8.1), and time intervals of predominant variants in Italy.

Fig. 8.1 shows that IFR_1 strongly increases and then strongly decreases during the first phase of the epidemic. After this transitory period, it stabilizes at values around 3.5%. A slow decrease to 2.2% follows, maybe due to vaccination, and this trend does not change notably when Alpha and Delta variants become prevalent. On the contrary, a fast decrease corresponds to the period when the Omicron variant becomes prevalent, terminating at 1.1%.

Fig. 8.2 reports IFR_2, calculated according to Eq. (8.2), and time intervals of predominant variants in Italy.

Fig. 8.2 shows that IFR_2 also presents a transitory period and then it stabilizes at values between 3% and 5%. A slow decrease to 2.6% follows, maybe due to vaccination, and this decrease does not change notably when Alpha and Delta

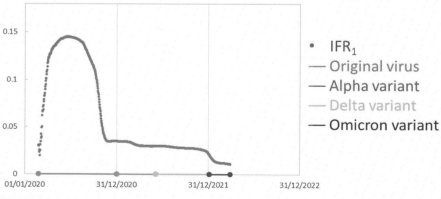

FIGURE 8.1

Infection fatality rate (IFR) calculated as cumulative deaths divided by cumulative confirmed positive cases, and time intervals of predominant variants in Italy.

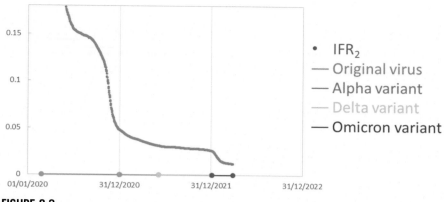

FIGURE 8.2

Infection fatality rate (IFR) calculated as cumulative deaths divided by cumulative resolved cases, and time intervals of predominant variants in Italy.

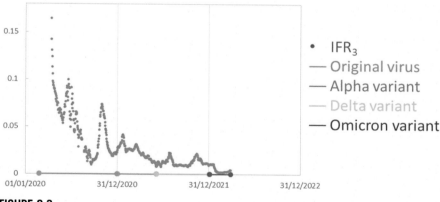

FIGURE 8.3

Infection fatality rate (IFR) calculated as daily deaths divided by positive cases confirmed 24 days before (average over the last 7 days), and time intervals of predominant variants in Italy.

variants become prevalent. On the contrary, a fast decrease corresponds to the period when the Omicron variant becomes prevalent, terminating at 1.2%.

Fig. 8.3 reports IFR_3, calculated according to Eq. (8.3) and averaged over the last 7 days to decrease stochastic variance, and time intervals of predominant variants in Italy.

Fig. 8.3 shows that IFR_3 also presents a transitory period at the first phase of infection. Moreover, it presents waves corresponding to infection waves. However, after the transitory, waves are slighter, and it stabilizes at values between 4.0% and 2.3%. A slowly decreasing trend to 1.0% follows, maybe due to vaccination, that does not change notably when Alpha and Delta variants become prevalent. On the contrary, a fast decrease to 0.2% corresponds to the period when Omicron variant becomes prevalent, terminating at 0.4%.

IFR_4 is also calculated, according to Eq. (8.5). While the same calculation in Ref. [34] indicated 0.9%, with data pertaining to the first infection period until October 2020, here it results in 0.2%, by taking into account some days in summer 2021, when a relatively high number of the test was performed.

8.4 Discussion and conclusion

The results reported above allow for making various observations.

From a qualitative point of view, the reported results show that:

- Results obtained are obviously different, if IFR is calculated as cumulative deaths divided by cumulative cases (IFR_1), as cumulative deaths divided by cumulative recovered (IFR_2), or as daily new deaths divided by new infections (IFR_3).
- IFR1 and IFR2 become more similar as the number of unresolved cases becomes more negligible.
- IFR_1 and IFR_2 estimate fatality on a cumulative basis; therefore, changes in these indices at some time points imply changes in the overall fatality until that time with updated daily data. On the contrary, IFR_3 weights the fatality of the most recent few days; therefore, changes in IFR_3 imply quite contemporary changes in fatality.
- For all indices, a transitory period can be detected during the first phase of pandemic. This is partially due to overestimation errors, during the development of tests, until a quite constant overall underestimation of positive cases is reached. Other real causes of the first transitory may be the spreading of infection among most of the subjects at risk of severe prognosis, and the tuning of management and care guidelines.
- After the transitory, estimations of IFR_1 and IFR_2 are quite feasible, since they are both overestimated, by the inverse of the ratio detected/actual positive cumulative cases, quite constant over time.
- IFR_3 keeps being variable over time, since its overestimation depends on the ratio detected/actual positive new daily cases, just depending on infection waves. Stochastic variance is mitigated since it is averaged on the last 7 days.
- A slowly decreasing trend can be identified for all indices, starting from December 2020, when vaccines became to be distributed in Italy.
- There are no noticeable changes in fatality among the first three virus variants that were predominant in Italy, that is, the original virus, the Alpha variant, and the Delta variant.
- There is a strong decrease in IFR, calculated in three different ways, between the previous predominant variants and the Omicron variant.

On the other hand, from a quantitative point of view, the reported results can be resumed as follows:

- All the numbers reported are estimates, obtained from data pertaining to Italy. Their generalization to other situations needs to verify the hypotheses stated in Section 8.2.6.
- The variability of estimates over time prejudices the exact calculation of fatality during the ongoing SARS-CoV-2 epidemic.
- The IFR is overestimated by most of these numbers, for the underestimation of positive cases. For this reason, they should be interpreted as the maximum values of the fatality. In particular, for most estimates, a maximum between 3% and 5% before vaccines can be individuated, decreasing to a maximum between 1% and 2.6% after vaccines distribution, for all variants except Omicron. For Omicron variant, a maximum IFR between 0.2% and 0.4% is estimated.
- The fatality in the period of the Omicron variant decreases by about two times, if IFR is calculated on a cumulative basis (IFR_1 and IFR_2), which takes into account all cases from the beginning of the pandemic. This means that, after the Omicron variant became prevalent, the fatality of SARS-CoV-2, calculated on all the cases up to now (comprising all the variants), is halved. On the other hand, the fatality decreases by about five times, if IFR_3 is calculated in the last few days. The latter finding is the most actual and means that the fatality of the Omicron variant is about five times lower than the previous predominant variants.
- As IFR_4 is calculated, it is less overestimated. It is around 0.2% for all variants except Omicron. A five times lower value is expected for the Omicron variant, that is, around 0.04%.

Concluding, the presented study allows evaluation (qualitatively and quantitatively, with some limitations) of the trend of fatality associated with SARS-CoV-2 variants. The results confirm previous studies [35] showing no overall differences among the original virus and Alpha and Delta variants and found their IFR estimate of about 0.2%, depending on vaccine distribution. On the contrary, according to previous findings [58], results show that death associated with the Omicron variant is less probable and found its IFR estimate about five times lower with respect to the previous predominant variants.

In order to monitor the importance of the global healthcare emergency, the decrease in fatality is good news. On the other hand, the total amount of deaths also depends on the infection rate, which increases as a new variant becomes predominant. Therefore, a promising research direction will include both infection rate and mortality, to find their correlation with both virus variants and vaccine distribution.

References

[1] World Health Organization. WHO Coronavirus Disease (COVID-19) Dashboard. [Internet]. 2020 [cited 2022 Apr 5]. Available from: https://covid19.who.int (accessed on 31 October 2020).

[2] S. Weston, M.B. Friedman, COVID-19: knowns, unknowns, and questions, mSphere 5 (2020) e00203−e00220.

[3] World Health Organization. Living guidance for clinical management of COVID-19. [Internet]. 2021 [cited 2022 Apr 5]. Available from: https://www.who.int/publications/i/item/WHO-2019-nCoV-clinical-2021-2.

[4] World Health Organization. Therapeutics and COVID-19: living guideline. [Internet]. 2022 [cited 2022 Apr 5]. Available from: https://www.who.int/publications/i/item/WHO-2019-nCoV-therapeutics-2022.2.

[5] C. Huang, Y. Wang, X. Li, L. Ren, J. Zhao, Y. Hu, et al., Clinical features of patients infected with 2019 novel coronavirus in Wuhan, China, Lancet 395 (2020) 497−506.

[6] N. Chen, M. Zhou, X. Dong, J. Qu, F. Gong, Y. Han, et al., Epidemiological and clinical characteristics of 99 cases of 2019 novel coronavirus pneumonia in Wuhan, China: a descriptive study, Lancet 395 (2020) 507−513.

[7] J.W. Tang, P.A. Tambyah, D.S.C. Hui, Emergence of a novel coronavirus causing respiratory illness from Wuhan, China J. Infect. 80 (2020) 350−371.

[8] P. Habibzadeh, E.K. Stoneman, The novel coronavirus: a bird's eye view, Int. J. Occup. Envrion. Med. 11 (2020) 65−71.

[9] World Health Organization, Global Surveillance for Human Infection with Novel Coronavirus (2019-nCoV), World Health Organization, Geneva, Switzerland, 2020.

[10] World Health Organization, Clinical Management of Severe Acute Respiratory Infection When Novel Coronavirus (2019-nCoV) Infection Is Suspected, World Health Organization, Geneva, Switzerland, 2020.

[11] C. Russell, J. Millar, J. Baillie, Clinical evidence does not support corticosteroid treatment for 2019-nCoV lung injury, Lancet 395 (2020) 473−475.

[12] C.I. Paules, H.D. Marston, A.S. Fauci, Coronavirus infections—more than just the common cold, JAMA 323 (2020) 707−708.

[13] M.L. Holshue, C. DeBolt, S. Lindquist, K.H. Lofy, J. Wiesman, H. Bruce, et al., First case of 2019 novel coronavirus in the United States, N. Engl. J. Med. 382 (2020) 929−936.

[14] Z. Wang, X. Chen, Y. Lu, F. Chen, W. Zhang, Clinical characteristics and therapeutic procedure for four cases with 2019 novel coronavirus pneumonia receiving combined Chinese and Western medicine treatment, Biosci. Trends 14 (2020) 64−68.

[15] M. Wang, R. Cao, L. Zhang, X. Yang, J. Liu, M. Xu, et al., Remdesivir and chloroquine effectively inhibit the recently emerged novel coronavirus (2019-nCoV) in vitro, Cell Res. 30 (2020). 369−271.

[16] E. de Wit, F. Feldmann, J. Cronin, R. Jordan, A. Okumura, T. Thomas, et al., Prophylactic and therapeutic remdesivir (GS-5734) treatment in the rhesus macaque model of MERS-CoV infection, Proc. Natl. Acad. Sci. U. S. A. 3 (2020) 20192208.

[17] X. Liu, X.-J. Wang, Potential inhibitors for 2019-nCoV coronavirus M protease from clinically approved medicines, J. Genet. Genom. 47 (2020) 119.

[18] H. Lu, Drug treatment options for the 2019-new coronavirus (2019-nCoV), Biosci. Trends 14 (2020) 69−71.

[19] World Health Organization. Status of COVID-19 Vaccines within WHO EUL/PQ evaluation process. [Internet]. 2022 [cited 2022 Apr 5]. Available from: https://extranet.who.int/pqweb/sites/default/files/documents/Status_COVID_VAX_02April2022.pdf.

[20] World Health Organization. Preventing and mitigating COVID-19 at work: policy brief, 19 May 2021. [Internet]. 2021 [cited 2022 Apr 5]. Available from: https://www.who.int/publications/i/item/WHO-2019-nCoV-workplace-actions-policy-brief-2021-1.

[21] World Health Organization. Tracking SARS-CoV-2 variants. [Internet]. 2022 [cited 2022 Apr 5]. Available from: https://www.who.int/en/activities/tracking-SARS-CoV-2-variants/.

[22] Presidenza del Consiglio dei Ministri - Dipartimento della Protezione Civile. pcm-Dpc/COVID-19. [Internet]. 2020 [cited 2022 Apr 5]. Available from: https://github.com/pcm-dpc/COVID-19.

[23] Istituto Superiore di Sanità. Press Releases. [Internet]. 2003 [cited 2022 Apr 5]. Available from: https://www.iss.it/en/comunicati-stampa.

[24] C. Wang, P.W. Horby, F.G. Hayden, G.F. Gao, A novel coronavirus outbreak of global health concern, Lancet 395 (2020) 470−473.

[25] World Health Organization. Novel Coronavirus Press Conference at United Nations of Geneva. [Internet]. 2020 [cited 2022 Apr 5]. Available from: https://www.who.int/docs/default-source/coronaviruse/transcripts/who-audio-script-ncov-rresser-unog-29jan2020.pdf?sfvrsn = a7158807_4.

[26] World Health Organization. Report of the WHO-China Joint Mission on Coronavirus Disease 2019 (COVID-19). [Internet]. 2020 [cited 2022 Apr 5]. Available from: https://www.who.int/docs/default-source/coronaviruse/who-china-joint-mission-on-covid-19-final-report.pdf.

[27] World Health Organization. WHO Director-General's opening remarks at the media briefing on COVID-19—3 March 2020. [Internet]. 2020 [cited 2022 Apr 5]. Available from: https://www.who.int/dg/speeches/detail/who-director-general-s-opening-remarks-at-the-media-briefing-on-covid-19-3-march-2020.

[28] A. Kazemi-Karyani, R. Safari-Faramani, S. Amini, V. Ramezani-Doroh, F. Berengjian, M.Y. Dizaj, et al., World one-hundred days after COVID-19 outbreak: incidence, case fatality rate, and trend, J. Educ. Health Prom. 9 (2020) 199.

[29] A.C. Ghani, C.A. Donnelly, D.R. Cox, J.T. Griffin, C. Fraser, T.H. Lam, et al., Methods for estimating the case fatality ratio for a novel, emerging infectious disease, Am. J. Epidemiol. 162 (2005) 479−486. Available from: https://doi.org/10.1093/aje/kwi230.

[30] Worldometer. Coronavirus (COVID-19) Mortality Rate. [Internet]. 2020 [cited 2022 Apr 5]. Available from: https://www.worldometers.info/coronavirus/coronavirus-death-rate/#nhc.

[31] Worldometer. COVID-19 Coronavirus Pandemic. [Internet]. 2020 [cited 2022 Apr 5]. Available from: https://www.worldometers.info/coronavirus/ (accessed on 31 October 2020).

[32] K. Mizumoto, K. Kagaya, G. Chowell, Early epidemiological assessment of the transmission potential and virulence of coronavirus disease 2019 (COVID-19) in Wuhan City, China, January-February, 2020, BMC Med. 18 (2020) 217.

[33] COVID-19 Forecasting Team, Variation in the COVID-19 infection−fatality ratio by age, time, and geography during the pre-vaccine era: a systematic analysis, Lancet (2022). Available from: https://doi.org/10.1016/S0140-6736(21)02867-1.

[34] M. Pota, A. Pota, M.L. Sirico, M. Esposito, SARS-CoV-2 infections and COVID-19 fatality: estimation of infection fatality ratio and current prevalence, Int. J. Environ. Res. Public Health 17 (24) (2020) 9290. Available from: https://doi.org/10.3390/ijerph17249290.

[35] M. Cetin, P.O. Balci, H. Sivgin, S. Cetin, A. Ulgen, H. Dörtok Demir, et al., Alpha variant (B.1.1.7) of SARS-CoV-2 increases fatality-rate for patients under age of 70 years and hospitalization risk overall, Acta Microbiol. Immunol. Hung. 68 (3) (2021) 153−161.

[36] Centers for Disease Control and Prevention COVID-19 Incidence and Death Rates Among Unvaccinated and Fully Vaccinated Adults with and Without Booster Doses During Periods of Delta and Omicron Variant Emergence—25 U.S. Jurisdictions, April 4—December 25, 2021. [Internet]. 2021 [cited 2022 Apr 5]. Available from: https://www.cdc.gov/mmwr/volumes/71/wr/mm7104e2.htm.

[37] World Health Organization. Tracking SARS-CoV-2 variants. [Internet]. 2020 [cited 2022 Apr 5]. Available from: https://www.who.int/en/activities/tracking-SARS-CoV-2-variants/

[38] F. Zhou, T. Yu, R. Du, G. Fan, Y. Liu, Z. Liu, et al., Clinical course and risk factors for mortality of adult inpatients with COVID-19 in Wihan, China: a retrospective cohort study, Lancet 395 (2020) 1054—1062.

[39] A. Woodruff, COVID-19 follow up testing, J. Infect. 81 (2020) 647—679.

[40] S.A. Lauer, K.H. Grantz, Q. Bi, F.K. Jones, Q. Zheng, H.R. Meredith, et al., The incubation period of coronavirus disease 2019 (COVID-19) from publicly reported confirmed cases: estimation and application, Ann. Intern. Med. 172 (2020) 577—582.

[41] NYC Health—Confirmed and Probable COVID-19 Deaths Daily Report. [Internet]. 2020 [cited 2022 Apr 5]. Available from: https://www1.nyc.gov/assets/doh/downloads/pdf/imm/covid-19-deaths-confirmed-probable-daily-05022020.pdf.

[42] New York City Department of Health and Mental Hygiene (DOHMH) COVID-19 Response Team, Preliminary Estimate of Excess Mortality During the COVID-19 Outbreak—New York City, March 11—May 2, 2020, MMWR Morb Mortal Wkly Rep 69 (2020) 603—605. http://doi.org/10.15585/mmwr.mm6919e5.

[43] C. Bulut, Y. Kato, Epidemiology of COVID-19, Turk. J. Med. Sci. 50 (2020) 563—570.

[44] New York State Governor. Amid Ongoing COVID-19 Pandemic, Governor Cuomo Announces Results of Completed Antibody Testing Study of 15,000 People Showing 12.3 Percent of Population Has COVID-19 Antibodies. [Internet]. 2020 [cited 2022 Apr 5]. Available from: https://www.governor.ny.gov/news/amid-ongoing-covid-19-pandemic-governor-cuomo-announces-results-completed-antibody-testing.

[45] United States Census Bureau. Quick Facts New York City, New York. [Internet]. 2020 [cited 2022 Apr 5]. Available from: https://www.census.gov/quickfacts/fact/table/newyorkcitynewyork/AGE775219#AGE775218.

[46] New York State Governor. NYC Health—Coronavirus Disease 2019 (COVID-19) Daily Data Summary. [Internet]. 2020 [cited 2022 Apr 5]. Available from: https://www1.nyc.gov/assets/doh/downloads/pdf/imm/covid-19-daily-data-summary-05022020-1.pdf.

[47] World Health Organization. Reducing public health risks associated with the sale of live wild animals of mammalian species in traditional food markets. [Internet]. 2021 [cited 2022 Apr 5]. Available from: https://www.who.int/publications/i/item/WHO-2019-nCoV-Food-safety-traditional-markets-2021.1.

[48] E.G. Price-Haywood, J. Burton, D. Fort, L. Seoane, Hospitalization and mortality among Black patients and White patients with Covid-19, N. Engl. J. Med. 38 (2020) 2534—2543.

[49] K. Khunti, L. Platt, A. Routen, K. Abbasi, Covid-19 and ethnic minorities: an urgent agenda for overdue action, BMJ 369 (2020) m2503.

[50] Y. Bai, L. Yao, T. Wei, F. Tian, D.Y. Jin, L. Chen, et al., Presumed asymptomatic carrier transmission of COVID-19, JAMA 323 (2020) 1406—1407.

[51] D. Wang, B. Hu, C. Hu, F. Zhu, X. Liu, J. Zhang, et al., Clinical characteristics of 138 hospitalized patients with 2019 novel coronavirus-infected pneumonia in Wuhan, China, JAMA 323 (2020) 1061–1069.

[52] L.M. Chang, L. Wei, L. Xie, G. Zhu, C.S.D. Dela Cruz, L. Sharma, Epidemiologic and clinical characteristics of novel coronavirus infections involving 13 patients outside Wuhan, China, JAMA 323 (2020) 1092–1093.

[53] Q. Li, X. Guan, P. Wu, X. Wang, L. Zhou, Y. Tong, et al., Early transmission dynamics in Wuhan, China, of novel coronavirus-infected pneumonia, N. Engl. J. Med. 382 (2020) 1199–1207.

[54] W. Liu, Q. Zhang, J. Chen, R. Xiang, H. Song, S. Shu, et al., Detection of Covid-19 in children in early January 2020 in Wuhan, China, N. Engl. J. Med. 382 (2020) 1370–1371.

[55] K.L. Shen, Y.H. Yang, Diagnosis and treatment of 2019 novel coronavirus infection in children: a pressing issue, World J. Pediatr. 16 (2020) 219–221.

[56] J.F.-W. Chan, S. Yuan, K.-H. Kok, K.K.-W. To, H. Chu, J. Yang, et al., A familial cluster of pneumonia associated with the 2019 novel coronavirus indicating person-to-person transmission: a study of a family cluster, Lancet 395 (2020) 514–523.

[57] H. Harapan, N. Itoh, A. Yufika, W. Winardi, S. Keam, H. Te, et al., Coronavirus disease 2019 (COVID-19): a literature review, J. Infect. Public Health 13 (2020) 667–673.

[58] Office for National Statistics. Comparing the risk of death involving coronavirus (COVID-19) by variant, England: December 2021. [Internet]. 2021 [cited 2022 Apr 5]. Available from: https://www.ons.gov.uk/peoplepopulationandcommunity/healthandsocialcare/causesofdeath/articles/comparingtheriskofdeathinvolvingcoronaviruscovid19byvariantengland/december2021.

Artificial intelligence for chest imaging against COVID-19: an insight into image segmentation methods

Rossana Buongiorno[1], Danila Germanese[1], Leonardo Colligiani[2], Salvatore Claudio Fanni[2], Chiara Romei[3] and Sara Colantonio[1]

[1]Institute of Information Science and Technologies "A. Faedo", National Research Council of Italy, Pisa, Italy
[2]Department of Translational Research, Academic Radiology, University of Pisa, Pisa, Italy
[3]2nd Radiology Unit, Pisa University Hospital, Pisa, Italy

9.1 Introduction

Coronavirus disease 2019 (COVID-19), caused by the severe acute respiratory syndrome virus 2 (SARS-CoV-2), has emerged as a pandemic leading to a global public health crisis. SARS-CoV-2 is mainly transmitted through respiratory droplets, and the infection may cause both mild symptoms of the upper respiratory tract as well as severe pneumonia. The disease typically presents with nonspecific symptoms, ranging from fever, cough, shortness of breath, and fatigue, to nausea, vomiting, and diarrhea. The clinical presentation of COVID-19 overlaps with that of other respiratory viral illnesses, and disease progression is quite rapid.

The incidence of COVID-19 has exerted, and still exerts, considerable pressure on healthcare resources, thus posing an urgent need for effective tools to support the diagnosis and clinical management of COVID-19 patients, as well as to increase the clinical efficiency of healthcare services.

In the early months of the pandemic outbreak, only the traditional RT-PCR method had the capability to detect the disease with accuracy, though requiring high detection time and reagents [1,2]. Nevertheless, there was an urgent need to provide fast and accurate diagnosis, in order to enact counteractions (i.e., isolation and tracking). In that period, multimodal chest imaging became a crucial tool for the early detection of COVID-19, and it served particularly the clinical management of the disease of symptomatic patients, since the imaging pathological findings basically reflect the inflammatory pathological process of the lung tissue [3].

Artificial Intelligence in Healthcare and COVID-19. DOI: https://doi.org/10.1016/B978-0-323-90531-2.00008-4

The clinical indications for imaging examinations have evolved during the pandemic's course, with diverse protocols adopted by the various national health services and clinical centers [4]. When the rapid antigen tests (nasal, oropharyngeal, and saliva swabs) were developed to enable the rapid diagnosis (even if not as accurate as RT-PCR test) for screening purposes, chest imaging served more to assess disease severity and progression [1,2,5], especially in severe patients [6,7].

The most commonly used imaging techniques include chest X-ray (CXR), lung ultrasound (LUS), and computed tomography (CT) [8,9], while the use of other imaging modalities, such as magnetic resonance imaging and positron emission tomography-CT, has been only rarely reported in the literature so far [4,7].

CXR has been widely used for fast diagnostic confirmation of COVID-19, thanks to its high availability, portability, and cost-effectiveness. LUS has been used for bedside evaluation and confirmation of the disease, as it is non-invasive, radiation-free, and highly portable [4,5]. CT imaging has been used to better distinguish the different parenchymal patterns and manifestations caused by the disease, thanks to its highest sensitivity and resolution [8,9].

Worth noting that detecting and quantifying such manifestations in terms of lung involvement is a key step to evaluate disease impact and to track its progression or regression over time. Nevertheless, the visual analysis and manual reporting by radiologists of CT examinations have revealed to be a challenging and time-consuming task, due to the high number of cases to report, the magnitude of the imaging data (e.g., a high-resolution CT volumetric acquisition comprises more than 100 slices), as well as the similarity of COVID-19 patterns with other types of pneumonia. During the peak of the pandemic spread, this task has put under pressure the radiology units all over the world, thus urging as never before the need for computer-aided diagnostic tools able to reduce radiologists' workload with a faster, consistent, and quantitative analysis of chest imaging data.

The scientific and clinical community has mobilized with a huge effort to respond to this need and to fight the overall pandemic crisis, also encouraged by funding bodies and research associations. This has resulted in a plethora of methods, research endeavors, and articles published in the literature. The building blocks of the proposed solutions leverage the most recent scientific and technological advances at the crossroads of artificial intelligence (AI), machine learning (ML), and computer vision (CV) applied to medical image understanding [7,10–14].

Several of the proposed AI-powered methods have addressed the extraction and quantification of the amount of pathological lung regions from the imaging data [7,10,11]. This is an important step in itself in the field of quantitative imaging [15], and it often serves as the starting point for further diagnostic and prognostic analyses. From a computational standpoint, this task is commonly addressed via the so-called *image segmentation* methods. Image segmentation aims to identify pixels belonging to the anatomical and pathological districts of interest, by distinguishing them from each other and the background. In COVID-19, the segmentation of chest imaging data responds to the need of detecting and differentiating the radiological manifestations of the lung parenchyma, by delimiting the corresponding image regions. This delimitation lays the basis for extracting quantitative information and textural parameters of those regions, to assess the lung volume involvement and the type of involvement.

This chapter provides the reader with an overview of the AI-powered methods that support the segmentation of COVID-19 chest imaging data. The focus is on chest CT examinations as CT better helps to discriminate the different parenchymal patterns that characterize COVID-19. A detailed description is provided for ML methods that leverage deep learning (DL) approach, as this has become in the last years the most commonly used approach, especially in the medical field.

As a starting point, Section 9.1 illustrates the radiological manifestations of COVID-19 and the main findings in CT imaging. Section 9.2 introduces the image segmentation task from a general perspective and the most common approaches to address it, along with their evaluation metrics. An overview of the methods that have been proposed in the literature to segment chest CT imaging in COVID-19 is reported in Section 9.3. A novel method that leverages innovative attention-based learning approaches is illustrated in Section 9.4. Finally, Section 9.5 discusses the potentials and challenges of image segmentation in COVID-19, by drawing the conclusions from the chapter.

9.2 Chest CT findings of COVID-19 pneumonia

With its high contrast and resolution, chest CT imaging plays a valuable role in distinguishing the different pathological parenchymal patterns and manifestations caused by COVID-19. CT scans allow for acquiring large volumes in a few seconds, decreasing the risk of artifacts due to the patient's respiratory movements. This modality ensures the absence of superimposed structures and the possibility to reconstruct the acquired images in different planes. CT has been recommended for thoracic evaluation in patients with moderate to severe symptoms or those experiencing respiratory functional impairment after recovery from infection [16,17]. For these patients, a CT examination provides a baseline evaluation for future comparison, it may suggest an alternative diagnosis, and it may establish the manifestations of important comorbidities in patients with risk factors for disease progression, thus guiding the treatment strategy [18].

The most common radiological patterns of SARS-CoV-2 infection is *ground-glass opacity* (GGO), particularly on the peripheral and lower lobes, and bilateral multiple lobular and subsegmental areas of *consolidation* (Fig. 9.1).

GGO refers to an area of increased hazy attenuation and density through which it is still possible to see vessels and bronchial structures. The pathological change of GGO is that the virus invades the bronchioles and alveolar epithelium, and replicates in the epithelial cells, causing the alveolar cavity to leak and the alveolar wall or the alveolar space to become inflamed or thickened, with distribution mainly around the lung and under the pleura. GGO may associate with other parenchymal alterations, such as interstitial thickening, generating the

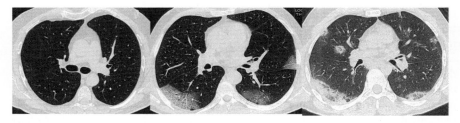

FIGURE 9.1

High-resolution computed tomography of normal lungs without GGO and consolidations (left); COVID-19 patient with GGO in the left lung (center); COVID-19 patient with consolidation in the right lung (right).

so-defined crazy paving pattern, or variably with consolidated parenchyma, thus generating the halo sign or the reversed halo sign.

Consolidation refers to a region of normally compressible lung tissue that has filled with liquid instead of air. The liquid can be pulmonary edema, inflammatory exudate, pus, inhaled water, or blood, and the condition is marked by the induration of a normally aerated lung.

Finally, nontypical imaging findings include pleural effusion, masses, cavitations, and lymphadenopathies; these would suggest alternative diagnoses [19].

Disease evolution varies sensibly from one patient to another, and there is no unique course of the disease. Nevertheless, in earlier stages, the main CT findings are most commonly GGO, and hazy regions with increased density with a subpleural involvement of lower lobes either unilaterally or bilaterally, while the indicators of the disease progression during the severe stage are multifocal, patchy, or segmental consolidations with more reticular configuration and thickened pulmonary interstitium [20−23]. The peak of the disease is characterized by acute respiratory distress syndrome, which consists in consolidations and areas with higher density involving almost entirely the lungs. As the disease progresses, the opacities may increase in size and density leading to diffuse opacity, consolidation, and thickened interlobular septa. Lung opacities may rapidly evolve into consolidations within 1−3 weeks of symptom onset, often peaking around 6−12 days after initial clinical presentation. When lung disease involves most of the lung parenchyma, patients may require intubation [2,24].

9.3 Medical image segmentation and artificial intelligence

Image segmentation plays a vital role in supporting diagnostic and prognostic processes, as it consists in identifying and selecting only the regions of interest in an image, slice, or volumetric scan, such as the organ of interest (e.g., the lungs) or the tissues or patterns of interest (e.g., GGO or consolidations).

Considering the image understanding pipeline (see Fig. 9.2), image segmentation is usually preceded by an image preprocessing step, which aims at enhancing image quality by removing noise and artifacts, harmonizing the data (e.g., normalizing or homogenizing them), and/or selecting specific parts (e.g., crops) of the data, to eliminate redundant information. Depending on the approach, segmentation may require a *feature extraction* step, which aims to extract parameters and information (e.g., related to the intensity of a pixel and its neighborhood) that may be relevant to distinguish the various regions of interest. Image segmentation is then followed by simple quantification steps (e.g., the estimation of the area or volume of an organ, a tumor, or a pattern) or more complex steps such as the extraction of morphological and textural features, which are further processed for diagnostic and prognostic purposes (see Fig. 9.2).

In the most advanced settings, image segmentation methods are completely automated, so as to avoid any burden posed to end users. Nevertheless, semiautomated techniques are also often delivered, which require verification, supervision, or input by an expert radiologist. Overall, a desirable and good solution should entail very few interactions [25].

The most common image segmentation techniques are usually categorized based on the image element they consider (i.e., intensity, edges, or regions) as well as the computational construct they employ (e.g., mathematical models, statistical or AI and ML approach). A deep analysis of the whole range of methods is out of the scope of this chapter and can be found in more extensive surveys, such as [26]. In the following, we overview briefly the classical segmentation

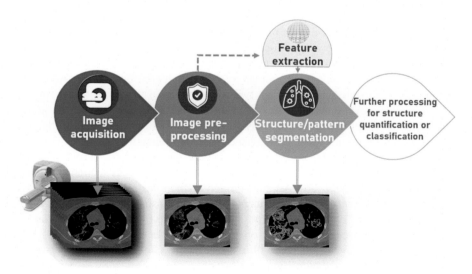

FIGURE 9.2

Image segmentation steps.

approaches and, then, focus on the most recent advanced techniques, which leverage a subfield of ML, known as deep learning (DL).

The most classical and common approaches that appeared in the literature until the early 2000s may be grouped into the following categories [26]:

- *Intensity- or histogram-based methods* exploit the distribution of the pixel intensity to separate the various regions of interest. Image thresholding methods, such as Otsu [27] and all the variations, are one of the most renowned methods of this group. Thresholding is well suited when there is a high contrast among the various regions and/or the background. In medical imaging, this condition is seldom fulfilled, and usually, thresholding serves as a first step followed by more complex algorithms.
- *Edge-based methods* exploit the spatial discontinuities in image intensity by applying simple edge detection operators (e.g., Sobel filter or the Canny operator) or transforms (e.g., Hough transform) [28]. Edge-based methods locate the edge within an image, thus requiring further processing to extract any region of interest. For this reason, they are usually combined with other methods to form a segmentation algorithm [29].
- *Region-based methods* work to delimit regions whose pixels share some form of similarity, usually based on intensity values. Region growing and watershed are the most renowned methods of this group. Region growing is very sensitive to noise, and it requires a manual interaction to obtain the seed points from which the regions grow. It produces often holed or disconnected regions or connected regions that should instead be separated. Usually, it requires a further post-processing step [30]. Watershed usually ensures a better outcome, but it may cause *oversegmentation*, by segmenting the image into an unnecessarily large number of regions [31].
- *Deformable models* leverage model-based parametric curves or surfaces and iteratively deform them under the influence of internal and external forces (based on intensity or discontinuities) to delineate region boundaries. Active contours and level-set methods are typical examples of this group [32]. These methods have experienced good success in the medical field and are sill quite used [33,34]. Nevertheless, they may suffer from poor convergence, in the case of concave boundaries, and depend on the initialization.
- *Atlas-guided or template-based methods* exploit the knowledge of anatomical structures depicted in medical atlases. These methods address segmentation as a registration problem [26]. An atlas-guided method has recently appeared in the literature for segmenting the lungs in magnetic resonance imaging [35].
- *Markov random field models* exploit Markov random fields to detect spatial interactions (between neighboring or nearby pixels) as part of other segmentation methods, such as within the K-means clustering [36] or active contour and watershed algorithms [37].
- *AI and ML-powered methods* address image segmentation task as pattern recognition task and, thus, process image-extracted features (usually based

on pixel intensities) to recognize whether a pixel belongs to the class of the regions of interest. These methods require a preliminary step for the extraction of image features (as shown in Fig. 9.2). Classification and clustering belong to this group, being the former based on a *supervised learning* approach and the latter on an *unsupervised learning* one. Supervised learning requires that, during the training phase, the classifier is provided with the desired output, namely, a ground-truth mask of the regions to be segmented. Unsupervised methods, instead, exploit some form of similarity measure to identify regions composed of similar pixels. Various models of artificial neural networks (ANNs) are among the most common approached within the supervised classification [38,39]. K-means and its variations are among the most common techniques adopted for unsupervised clustering [40]. In some cases, clustering and classification have been used in combination [41,42].

All the above methods have experienced alternated successes and failures in medical image segmentation. Some authors have grouped them into three successive generations [43]. Nevertheless, we argue that, in the last decade, a fourth generation of methods has emerged (see Fig. 9.3).

This new generation leverages the most recent developments in the field of AI and ML. Actually, the recent deluge of data has boosted the advances of a particular type of ANN, the so-called convolution neural network (CNN), which has been demonstrated to perform unprecedentedly well, especially when solving perception tasks, such as vision, object recognition, and natural language processing [44]. Considering the depth (in terms of number of layers) of this type of network, the term *deep learning* has emerged when referring to them, thus standing out in opposition to the *shallow learning*, which is the traditional learning approach of ML methods and ANNs.

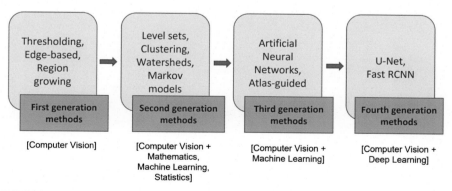

FIGURE 9.3

The four generations of image segmentation methods.

9.3.1 The fourth generation of segmentation methods: deep learning approaches

The most renowned DL model is undoubtedly CNN. Nevertheless, as better explained in the following, CNNs do not perfectly suit the segmentation task, and they have been modified for the scope. For the sake of clarity, we start providing the reader with a very brief introduction to CNNs and then move to illustrate the other models deriving from them that better address the segmentation task.

Convolutional neural networks: A CNN consists of several layers of connected and parallel processing units [44], which are able to take in input the whole image. The layers might be mainly of three types: *convolutional*, *pooling*, or *fully connected* layers (see Fig. 9.4).

The convolutional layers apply to the input data matrix a convolution operation with *feature maps* (or *kernels*) able to detect distinct elements or characteristics of the input, such as edges or other visual patterns. An *activation function*, such as the logistic or rectified linear unit function [44], selects the most relevant of these elements. The pooling layers are located after a convolutional layer to reduce the size of the data before they enter the following layer. Convolutional and pooling layers enable the input image to be transformed in increasing levels of abstraction. The fully connected layers are usually located at the end of the network and correspond to a classical ANN (i.e., a multilayer perceptron).

CNNs are trained by using an iterative error minimization process, which aims to minimize a loss function. The loss function depends on the problem at the end (i.e., classification or regression) and the learning approach employed (i.e., supervised or semi-supervised). Training a CNN might be a very demanding task, as there are no general rules to determine the most suitable architecture (e.g, number and type of layers, number and type of connections), training parameters (e.g., loss function, error minimization strategies), and optimization strategies (e.g., regularization based on dropout or weight decay [45]) that may guarantee the best results. Optimization strategies are particularly important to avoid the so-called *overfitting* phenomenon, which happens when the network has learned the

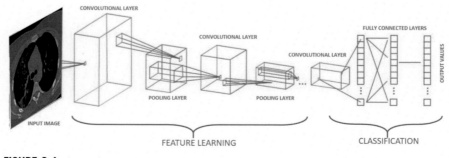

FIGURE 9.4

A sample CNN architecture made of convolutional, pooling, and fully connected layers.

training data, and it becomes highly specialized to recognize them, while it generalizes very badly on unseen data. Usually, developers undertake a *trial-and-error* process to figure the best solution out based on the data available.

CNNs usually serve supervised classification or regression purposes [46]. In image segmentation, CNNs have been initially employed for a pixel-based classification (also known as *dense prediction or dense classification* task), by processing mainly image patches [47,48]. Later on, considering that the fully connected layers cannot manage inputs and outputs that vary in size as it may happen in segmentation, variations of the standard CNN architecture have been introduced, demonstrating to suit more the segmentation purposes.

Two categories among the most well-known variations are *fully convolutional* NN (FCNN) and *region* CNN (see Fig. 9.5). In the following, we provide the reader with a brief overview of these two categories. For a more comprehensive classification and description, the reader is referred to dedicated survey papers [49,50].

Fully convolutional neural networks: An FCNN consists of only convolutional layers, meaning that the last fully connected part is replaced by additional convolutional layers (see Fig. 9.6) [51].

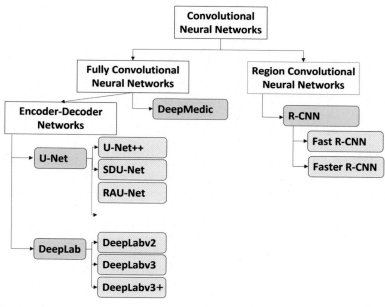

FIGURE 9.5

Some of the most renowned state-of-the-art DL models for image segmentation. *White* rectangular boxes indicate the network category; the *gray* rounded boxes are the DL models; and the rounded strip-filled boxes are the variations of such models.

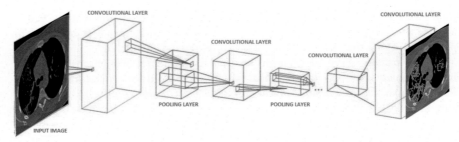

FIGURE 9.6

A sample FCNN architecture.

In the medical field, one of the first successful attempts has been DeepMedic [52], which relies on a 3D CNN for a dense classification whose results are then refined with fully connected *conditional random fields* [53].

FCNNs are at the basis of the *encoder–decoder* models, which have demonstrated very good performances in image segmentation. An encoder–decoder model encompasses two processing steps to map the input image to its corresponding output (i.e., the segmentation mask). The first step employs a set of convolutional layers that map or *encode* the input image into a latent space of features. The second steps leverage another set of convolutional layers to deconvolute or *decode* the *latent representation* into the expected output (i.e., the segmentation mask).

The most known encoder–decoder FCNN is undoubtedly U-Net [54], whose name derives from the U shape formed by its downsampling encoder and the upsampling decoding parts (see Fig. 9.7). The U-Net has been extensively used in medical image segmentation, both for 2D and 3D segmentation, as it is able to guarantee good results even if trained with a limited number of examples.

To date, several variants of the U-Net have been proposed and applied to medical image segmentation, such as U-Net++ [55], SD-UNet [56], and RA-UNet [57]. Among the different variants, nnU-Net (which stands for no-new-U-Net) has lately gained much interest, thanks to its general-purpose usage and self-configuring abilities [58]. Actually, designing and training a DL model may be very laborious and may require a considerable number of attempts to obtain sensible results. nnU-Net is particularly valuable as it automatically configures itself for any new segmentation task, also with respect to the data preprocessing, the network architecture, network training, and post-processing. nnU-Net has been applied also to respond to COVID-19 segmentation challenges [59].

DeepLab is another well-known state-of-the-art encoder–decoder FCC for image segmentation [60]. DeepLab uses *atrous* (with holes) convolution, to convolute the input data with upsampled filters, thus guaranteeing a wider field of view at the same computational cost. Several upgrades of DeepLab have appeared in the literature, such as DeepLabv2 [60,61], DeepLabv3 [62], and DeepLabv3 + [63], with diverse variations in the use of the atrous convolutions. In the medical

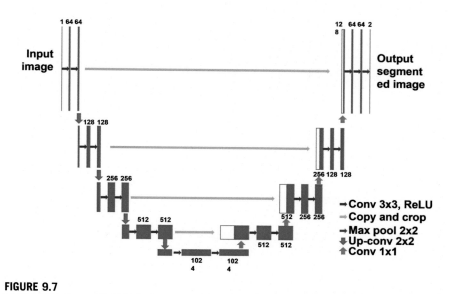

FIGURE 9.7

The U-Net architecture.

Inspired by the original paper O. Ronneberger, P. Fischer, T. Brox, U-net: Convolutional networks for biomedical image segmentation, in: Navab N, Hornegger J, Wells WM, Frangi AF, editors, Medical Image Computing and Computer-Assisted Intervention — MICCAI 2015, Springer International Publishing.

field, DeepLabv3++ has recently served tumor segmentation in diverse applications [64,65].

Region CNNs: The first region-based model proposed for object detection and segmentation task is the R-CNN [66]. Such a model generates region proposals as bounding boxes, based on a selective search process. These region proposals are then warped to standard squares and passed as input to a CNN for the extraction of feature maps. These features are then fed to the classification algorithm so as to classify the objects lying within the region proposal network (see Fig. 9.8).

As R-CNN iterates several times on each region proposal, it may require a long time to be trained. Several variants have been proposed to overcome this limitation, such as Fast R-CNN [67] and Faster R-CNN [68].

DL models have been demonstrated to work unprecedentedly well in image segmentation. Nevertheless, they present two important issues: data and computation eagerness. The training process serves to set the values of a huge number of internal parameters of the DL model (i.e., the more the layers, the more the parameters). This requires to expose the model to a commensurate amount of training data (i.e., large training sets) and to iterate the process many times (i.e., high computational power). The data availability issue is rather critical when

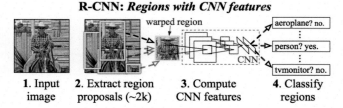

FIGURE 9.8

The processing workflow behind R-CNN.

Inspired by the original paper R. Girshick, J. Donahue, T. Darrell, J. Malik, Rich feature hierarchies for accurate object detection and semantic segmentation, in: 2014 IEEE Conference on Computer Vision and Pattern Recognition, 2014, pp. 580–587. https://doi.org/10.1109/CVPR.2014.81.

considering image segmentation, as this task is usually addressed by leveraging a supervised learning approach and, thus, requires the availability of high-quality annotated and labeled data. In medical imaging, this criticality appears even more challenging, as data annotation should come from expert clinicians and radiologists who manually draw the segmentation ground-truth masks. The data availability issue is often counteracted with *transfer learning*. Transfer learning consists in the reuse of knowledge and latent features learned from a network trained on other (larger) datasets. There are two kinds of widely adopted transfer learning strategies: network backbone transfer and network pretraining. The former strategy consists in borrowing a network trained with large-scale datasets and embedding it in another model as a backbone for extracting the latent features, which are then processed to accomplish the segmentation task at hand. The latter strategy consists in pretraining the network at hand using large-scale datasets and then performing formal training with the target dataset.

To date, U-Net and its variants have been the dominant model adopted or modified to address the segmentation of COVID-19 chest CT, as it will be extensively analyzed in Section 9.3. Fast R-CNNs have been mainly used for detection and segmentation in CXR imaging data [69]. Several approaches based on backbone transfer learning, pretraining, and the definition of novel architecture trained from scratch have appeared in the literature in the last couple of years. Before diving into detail about the solutions proposed for COVID-19, we briefly illustrate the evaluation metrics that serve assessing the performance of segmentation methods.

9.3.2 Evaluation metrics

Evaluating the performance of a segmentation method entails comparing its results against the expected ones. This corresponds to measuring the distance or similarity between two diverse segmentations: one is the segmentation to be evaluated, and the other is the corresponding ground-truth segmentation.

We report below some of the most common metrics for segmentation evaluation [70,71]:

- Pixel accuracy (PA)

    Pixel accuracy is the percent of pixels in the segmented image that are correctly classified. Let n_{ij} be the number of pixels of class i predicted to belong to class j, where there are ncl different classes, and let $\sum_j n_{ij} = t_i$ be the total number of pixels of class i, and we can define the pixel accuracy as (9.1):

$$PA = \frac{\sum_i n_{ii}}{\sum_i t_i} \tag{9.1}$$

    Unfortunately, high PA does not always imply superior segmentation ability, especially when the classes are extremely imbalanced, namely, a class or a subset of classes that dominate the image, while some other classes make up only a small portion of the image.

- *Dice score (D)*

    Dice score compares the segmentation mask predicted by the segmentation method (P) with the ground-truth mask (G), according to Eq. (9.2):

$$D(P, G) = \frac{2|P \cap G|}{|P| + |G|} \tag{9.2}$$

    D corresponds to the double area of overlap divided by the total number of pixels in both segmentations. The Dice score ranges from 0 to 1, with 1 signifying the greatest similarity between the predicted and the truth. When used in DL models, the Dice score may be used to define the loss function to be maximized.

- *Intersection over Union or Jaccard index*

    Extremely effective, Intersection over Union (IoU) is the area of overlap between the predicted segmentation and the ground truth divided by the area of union between the predicted segmentation and the ground truth, as in (9.3). This metric ranges from 0 to 1 (0%$-$100%) with 0 signifying no overlap and 1 signifying perfectly overlapping segmentation. For binary (two classes) or multi-class segmentation, the mean IoU of the image is calculated by taking the IoU of each class and averaging them.

$$JI(P, G) = \frac{|P \cap G|}{|P \cup G|} \tag{9.3}$$

9.4 Existing methods for COVID-19 chest CT images segmentation

Chest CT image segmentation of COVID-19 patients mainly encompasses identifying GGO and consolidations in the lungs. As introduced in Section 9.1, these are the CT imaging findings useful to assess the progression of lung involvement and the severity of the disease.

The methods presented in the literature addressed the problem by focusing on two main tasks, according to which they can be divided into two categories: the *lung-region-oriented* and the *lung-lesion-oriented* methods.

The *lung-region-oriented* methods aim to basically separate lung regions from the rest of the image. This type of segmentation supports clinicians in their visual inspection of CT imaging, being particularly useful to visually detect the edge of the lungs. *Lung-lesion-oriented* methods aim to separate the COVID-19 lesions within the lungs by providing either

- a binary segmentation: discriminating between the diseased and the healthy tissue or
- a multi-class segmentation: differentiating GGO, consolidations, and healthy tissue.

Worth noting that the binary segmentation differentiates the diseased tissue from the healthy one, thus offering the possibility to assess and quantify the overall lung involvement. Multi-class COVID-19 lesion segmentation separated different types of lesions simultaneously, thus enabling a more precise assessment and quantification of disease severity and progression [72]. One might, then, identify a hierarchy of more complex and comprehensive methods, which starts with lung-region-oriented methods, passes through the binary lung-lesion-oriented methods, and ends up with the multi-class lung-lesion-oriented ones.

For both lung-region-oriented and lung-lesion-oriented methods, the quantity and quality of available data highly influence the choice of the DL model to adopt. In this respect, to sustain the research in the field, several datasets have been made publicly available (as reported in Table 9.1).

Nevertheless, since these datasets appeared at various times in the past 2 years and are somewhat limited, the solutions proposed so far have often adopted a transfer learning approach, as illustrated in Fig. 9.9, which reports the diverse categories that group the segmentation methods that appeared to date.

Table 9.1 A summary of COVID-19 publicly available imaging datasets.

Dataset	Mode	N. total of image slices	N. COVID	Task
MosMedData [73]	CT	785 (50 cases)	785	Lesion segmentation (binary)
COVID-19 CT segmentation dataset [74]	CT	100 (50 cases)	100	Lesion and lung segmentation (multi-class)
COVID-19 Lung and Infection Seg. Data [75]	CT	1844 (20 cases)	1844	Lesion and lung segmentation (binary)
BIMCV COVID-19 + [76]	CT/ X-ray	5381 (1311 cases)	2428	Lesion and lung segmentation (binary)

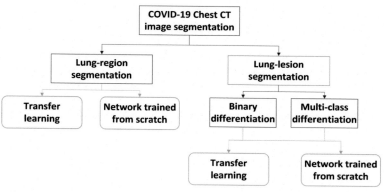

FIGURE 9.9

A categorization of COVID-19 CT segmentation DL methods. Rectangular boxes indicate the task accomplished, *while* the rounded boxes indicate the computational approach adopted to train the method.

In the following sections, we overview the various solutions by distinguishing between lung region and lung lesion approaches.

9.4.1 Lung-region-oriented methods

For the lung region segmentation in CT images, Maiello et al. proposed a 3D transfer learning approach to quantify lung volumes in COVID-19 patients [77]. The authors trained two U-Nets with different resolutions and fields of view. The first, called u2NetPig, was trained and evaluated on 180 scans of 18 pigs with experimental acute respiratory distress syndrome (ARDS). The second, u2Net$_{Human}$, was trained and tested on a clinical dataset of 150 scans from 58 intensive care unit patients with lung conditions varying from healthy, to chronic obstructive pulmonary disease, to ARDS and COVID-19. A manual segmentation was available for each scan, being a consensus by two experts. Transfer learning was then applied to train u2NetPig on the clinical dataset generating u2Net$_{Transfer}$. The quality of the segmentation was evaluated using the Dice score and the Jaccard index. u2Net$_{Human}$ reached a value of Dice score of 93.7% and of Jaccard index of 88.2%, while u2Net$_{Transfer}$ achieved 95.7% of Dice score and 91.8% of Jaccard index. The authors then demonstrated that transfer learning improved the overall performances.

As far as training from scratch is concerned, Tilborghs et al. [78] implemented a 3D neural network based on DeepMedic architecture [52], and they trained it from scratch to separate the lung regions. The training set contained 74 patients, 42 of which with a positive RT-PCR test, 24 with a negative RT-PCR test, and the remaining with an unknown status. The test set contained 7 patients with suspected but unconfirmed COVID-19 status and was enlarged with 10 extra

COVID-19 confirmed cases from a public dataset [75]. On all the images and the corresponding ground truths, the authors applied data augmentation by adding Gaussian noise and performing random affine transformations. The network was trained using a patch-based approach to predict segments of (43, 43, 13) voxels simultaneously. On the test set, the achieved value of the Dice score was 97.4%.

In addition, for the segmentation of lung lobes, the authors trained from scratch an ensemble of three 3D CNNs, with the same 3D architecture based on DeepMedic. Compared to the previous case, each CNNs had some minor changes: the final prediction layer and the convolutional kernel sizes. For this task, they used the publicly available LUNA16 dataset [79] from which 40 patients were randomly selected for training and test sets. LUNA16 dataset is a dataset for lung segmentation which consists of 1.186 lung nodules annotated in 888 CT scans. To obtain the ground truth, they used the online lobe segmentation tool of Xie et al. [80]. The authors applied data augmentation by adding Gaussian noise and performing random affine transformations. The networks were trained using a patch-based approach to predict segments of (43, 43, 13), (43, 13, 43), or (13, 43, 43) voxels simultaneously. On the test set, the achieved value of the Dice score was 97.6%.

9.4.2 Lung-lesion-oriented methods

9.4.2.1 Binary lung lesion methods

To date, most of the lung lesion segmentation methods have addressed the binary differentiation between the diseased and healthy tissue.

As far as the transfer learning approach is adopted, Wang et al. explored the benefits of transferring knowledge from CT images of non-COVID-19 patients and learning from multiple lung lesion datasets to extract more general features [81]. By using 20 CT scans from a publicly available COVID-19 CT dataset (i.e., the COVID-19 Lung and Infection Seg. Data [75]) and 191 CT scans from three public non-COVID-19 datasets [82−84], the authors presented a new transfer learning strategy using a standard 3D U-Net as the backbone network. They proposed a *hybrid-encoder learning* approach. It consisted in equipping the standard 3D U-Net with two encoders having the same architecture: a dedicated encoder and an adapted encoder. The dedicated encoder was a task-specific feature extractor, focusing on segmenting COVID-19 infection through re-initializing all the parameters. The adapted encoder was an auxiliary feature extractor, aiming to learn general lung lesion features from non-COVID-19 infection. Furthermore, to aggregate information from both the encoders, they developed an attention-based selective fusion unit in order to obtain a global and comprehensive representation for the decoding phase [85]. The results showed that the proposed hybrid-encoder strategy based on multi-lesion pre-trained model achieved a mean Dice score of 70.4%.

Still exploiting transfer learning, Liu proposed a novel two-stage cross-domain transfer learning framework of a DL model called nCovSegNet [86]. The first stage took advantage of a training session on ImageNet [87] and provided valuable cross-domain knowledge from the human perception of natural images. ImageNet is a large visual database designed for use in visual object recognition research, with more than 14 million hand-annotated natural images. Since a large gap exists between natural images and COVID-19 CT images, they further performed a second-stage transfer learning, where the model was pretrained using LIDC-IDRI [88], which is currently the largest CT dataset for pulmonary nodule detection and provides a large amount of chest CT images that share similar appearances with the COVID-19 CT images. Finally, they trained their model with 40 cases of COVID-19 from the MosMedData [73]. They applied data augmentation operations such as random horizontal flip, random crop, and multiscale resizing with different ratios (0.75, 1, 1.25). The performances were evaluated on a test set consisting in 10 CT scans of COVID-19 patients from MosMedData showing a Dice score of 68.43%. The authors compared the performances of nCovSegNet with those of other state-of-the-art models, like U-Net and U-Net++, which reached a value of Dice score, on the same test set, of 54.19% and 54.54%, respectively.

As far as training from scratch is concerned, Chassagnon et al. proposed CovidENet [89]: an ensemble of 2D and 3D CNNs based on AtlasNet [90]. AtlasNet combines CNN ensembling and spatial normalization by registration to a template. The authors used a dataset composed by 180 CT scans acquired in six different centers with four different CT models. The training set was composed of 50 CT scans from the first three centers, while the test set contained 130 CT scans from the remaining three centers. The authors predetermined the proportion between the CT manufacturers in the datasets in order to maximize the model generalizability while taking into account the data distribution. Two experts annotated the data: the patients from the training cohort were annotated slice-by-slice, while the patients from the testing cohort were partially annotated on the basis of 20 slices per examination covering in an equidistant manner the lung regions. The proposed method borrowed elements from already established FCNN architectures [91,92], and it incorporated powerful design aspects, such as deformable registration methods. Each CNN was associated with a template CT scan, such that all CT scans were registered to this template before being given as input to the network. The 2D architecture allowed to explore the spatial resolution on the axial space after mapping to a common space, while the integration of 3D networks was dictated by the interest of integrating consistency on the coronal/sagittal planes. The evaluation metric for the performance of the model was the Dice score which reached a mean value of 70% between CovidENet and the manual segmentations of the two experts on the test set.

Chaganti et al. trained from scratch a novel network [93]. They proposed to automatically segment GGO and areas of consolidation together, to provide a binary segmentation using DenseU-net [94]. The aim was to compute the percentage of opacity (PO) and lung severity score (LSS), as they quantify the extent of lung involvement and the distribution of involvement across lobes, respectively.

The authors used a combination of COVID-19, viral pneumonia, and other interstitial lung diseases (ILD) datasets, since these diseases have similar CT abnormalities as COVID-19, such as GGO and consolidation. Thus, the training set contained a total of 901 CT scans (431 COVID-19, 174 viral pneumonia, and 296 ILD). The test set consisted in 200 CT volumes belonging to the same dataset of the training set. The annotation task was performed manually using ITK-Snap by a team of 25 expert medical annotation engineers and was performed under the supervision of 3 board-certified radiologists. The authors trained a DenseU-Net with anisotropic kernels to transfer a 3D chest CT volume to a semantic segmentation mask of the same size. They used a single label to define all voxels within the lungs that fully or partially contain GGO or consolidations as positive voxels, while the rest of the image areas within the lungs and the entire area outside the lungs were defined as negative. The network was trained as an end-to-end segmentation system. Differently from the previous works, in which the authors have used the Dice score, the performances of the model proposed by Chaganti et al. were evaluated by using the Pearson correlation coefficient. The Pearson coefficient is used to present the linear correlation between random variables belonging to two groups of continuous data that maintain a linear relationship. Each pair of measured values is independent of each other. The Pearson coefficient takes values in between -1 and 1. In the work by Chaganti et al., it was calculated between the method's prediction and the ground truth for COVID-19 positive scans, and it reached a value of 0.92 for PO and 0.91 for LSS.

At the end of 2020, a MICCAI-endorsed challenge on COVID-19 segmentation boosted the development of various solutions for the binary segmentation of CT scans [59]. This challenge utilized data from two public resources of chest CT images, namely, the CT Images in COVID-19 [95] (Dataset 1) and COVID-19-AR [96] (Dataset 2), both available on The Cancer Imaging Archive (TCIA). The first dataset originated from China, while the second was from the United States. Only a subset of the first dataset was used to provide the challenge participants with a training and validation set (i.e., 199 cases for training and 50 for validation). Additional 46 cases, 23 borrowed from Dataset 1 and 23 from Dataset 2, were kept by the organizers for the test phase and not directly disclosed to the participants. Currently, these datasets are not openly available; for this reason, they have not been included in Table 9.1 [59].

All top 10 teams used a 2D/3D U-Net variant with only minor modifications [59]. Worth noting that five out of these teams used the nnU-Net [58]. The nnU-Net authors themselves ranked second in the final leaderboard. They leveraged the nnU-net framework to design five 3D U-Net configurations, varying the 3D patch size (i.e., patches of $28 \times 256 \times 256$ voxels or $40 \times 224 \times 192$ voxels), the use of data augmentation, and the use of batch or instance normalization. These five configurations were trained ten times, and the resulting 50 models were combined via a softmax averaging into a segmentation ensemble. Among the various models, the best-performing model was a single network (i.e., a high-resolution U-Net with extensive data augmentation and instance normalization), which obtained an average Dice score of 75.43% on the training set (only slightly higher than the baseline results of 74.41%) and 65.43% on the test set.

The top-ranked team exploited the entire original Dataset 1 from TCIA, though it was not labeled for the lesion segmentation task. The goal of the team was to improve the generalization ability of the segmentation model by using more training data. They leveraged a simple semi-supervised approach for training a nnU-Net. First, they trained the network using labeled data provided by the challenge organizers. Then, they utilized the trained segmentation model to generate the *pseudo*-lesion masks of both labeled and unlabeled CT images. This dataset with pseudo labels was used to train again the network in a fully supervised manner. This way, the team was able to obtain a Dice score equal to 76.65% on the training set and 66.59% on the test set.

For the other solutions, the reader is referred to the summary paper [59], as none of the top 10 models have been to date published in a dedicated paper.

As it is evident from the final performance scores, the challenge was particular hard, as the datasets included very heterogeneous cases. Moreover, each case was annotated by one radiologist who refined the results provided by a publicly available COVID lesion segmentation AI model [97]. A second verification from another human expert was then missing.

9.4.2.2 Multi-class lung lesion methods

Very few attempts have been done so far to address the multi-class lung lesion segmentation. We report here the work by Fan et al. who proposed a novel deep network, called Inf-Net [98].

The authors followed a two-step approach to develop Inf-Net. They used a 2D FCNN encoder—decoder architecture using reverse attention modules for the first step that aimed at a binary lesion segmentation. The second step consisted of an ensemble of two models: Inf-Net and FCN8 [99], and Inf-Net and U-Net. Both FCN8 and U-Net took the CT and the binary segmentation predicted in the first step as input. To train the network of the first step, the authors used both labeled and unlabeled data. They gather together a semi-annotated COVID-19 infection segmentation dataset, which consisted of 20 unlabeled CT volumes from the COVID-19 CT Collection [100] and 100 labeled CT volumes belonging to the COVID-19 CT Segmentation dataset [74]. They used this semi-annotated dataset to train a *semi-supervised* version of Inf-Net. This was done by using 45 CT labeled scans to first train the semi-supervised Inf-Net and gradually fine-tuning the model by ingesting in the training set new cases taken from the unlabeled dataset whose ground truth was generated by the same model (i.e., *pseudo*-ground-truth labels).

The authors extended Inf-Net to a multi-class lung infection labeling framework to distinguish between GGO and consolidation. Namely, the infection segmentation results provided by Inf-Net guided the multi-class labeling, as, in combination with the corresponding CT images, they were fed to a multi-class segmentation network, i.e., FCN8 and U-Net.

To compare the infection segmentation performance, the authors considered the two state-of-the-art models U-Net and U-Net++. The proposed Inf-Net outperformed U-Net and U-Net++ in terms of Dice score by reaching a value of 73.9% against 43.9% and 58.1%, respectively. The authors attributed this improvement to the reverse attention and explicit edge-attention modeling, which provided robust feature representations.

9.5 Attention-FCNN: a novel DL model for the segmentation of COVID-19 chest CT scans

Within the Tuscany Region project OPTMISED ("An optimized path for the clinical management of COVID-19 patients"), a multidisciplinary team is currently working to define novel AI and statistical models able to stratify COVID-19 patients and predict their outcome based on clinical, biochemical, and imaging data. Chest CT imaging is considered in the project to estimate the lung involvement and to extract quantitative biomarkers that may be relevant in outcome prediction. As a preliminary step in this respect, we designed, trained, and tested a 2D FCNN model for the binary segmentation of chest CT imaging data. The model leverages an attention-based learning approach into an encoder–decoder architecture; hence, we named it Attention-FCNN. The network was preliminary trained and tested on a dataset of 25,013 images, derived from 56 CT scans, retrospectively collected during the first period of the COVID-19 outbreak in one of the clinical centers involved in the project. Herein, we report in more detail the experiment done so far, with the aim of illustrating our research activities in the field. In the current release of the segmentation model, we took advantage of a preliminary annotation process of the dataset by the radiologists who isolated the lung pathological regions. The project activities, currently ongoing, are focusing on the delivery of multi-class annotations of the imaging data by expert radiologists. This will enable the design and development of more advanced DL segmentation model able to deliver a multi-class segmentation.

The following sections introduce the data used for the experiment (Section 9.4.1), the FCNN model architecture (Section 9.4.2), its training process, and the results obtained (Sections 9.4.3 and 9.4.4). To assess the contribution of the novel-introduced attention learning mechanism, we carried out an ablation study and report the result in Section 9.4.5.

9.5.1 Chest CT imaging dataset

A dataset of 56 HRCT volumetric scans was gathered from the 2nd Radiology Unit database of the Pisa University Hospital, belonging to 56 patients diagnosed with COVID-19. Each scan had about 250 slices with 512×512 pixels per slice, and each slice had the same pixel spacing (i.e., 0.7 mm). All the scans were

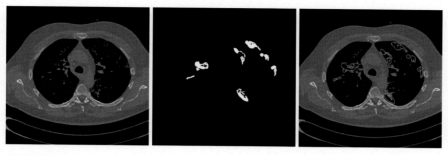

FIGURE 9.10

On the left: an example of the input original image; in the middle: the binary mask related to the segmentation of the infection in the original image used as ground truth during the training; on the right: the ground truth superimposed on the original image.

acquired using the same CT scanner (i.e., a Siemens Sensation 64), by following the same acquisition protocol.

The reference segmentation masks, used as ground truth, were obtained as an agreement among three expert radiologists, two of whom with a 5-year experience and one with more than 10-year experience. Preliminary segmentation masks were provided by using UIP-net [101]. UIP-net is an encoder–decoder convolutional neural network trained from scratch for the segmentation of the typical radiological findings of idiopathic pulmonary fibrosis (IPF). Since both IPF and COVID-19 manifest with GGO in chest CTs, the segmentation provided by UIP-net was used as a preliminary mask to facilitate radiologists' manual work. Once obtained the UIP-net masks, the two youngest radiologists proceeded by manually refining the results and adding consolidations to obtain the final segmentation masks, which were, then, reviewed and finalized by the eldest radiologist (see Fig. 9.10).

9.5.2 Attention-FCNN architecture

The Attention-FCNN was defined to take as input 2D axial images with 512×512 pixels. An encoder–decoder model, inspired by the U-Net, was selected as the basic architecture. It consists in a downsampling path in which the high-level features of the images are enhanced, and an upsampling path that reconstructs a pixel-wise segmentation from the high-level features. Skip connections [54] pass the features from the encoder to the decoder path to recover spatial information lost during downsampling. More precisely, the overall architecture is shown in Fig. 9.11.

The depth of both the encoder and the decoder path was set equal to 3. Each convolutional layer had a ReLU activation function and a receptive field of 3×3 pixels. To prevent overfitting, a random dropout regularization was used, with a 20% frequency, after each convolutional layer. Along the downsampling path, each convolutional layer (named d^i, $i = 1,2,3$) doubled the number of feature

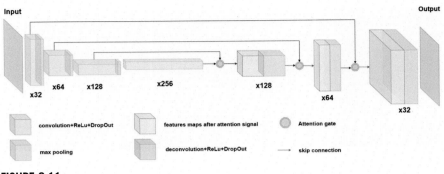

FIGURE 9.11

Attention-FCN architecture.

maps, starting from a stack of 32 and ending with a stack of 256 maps (in the apex layer p). Max-pooling layers were used to combine the convolutions. Along the upsampling path, starting from the 256 maps, each convolutional layer (named u^i, $i = 1,2,3$) halved the number of feature maps. No pooling layers were included. The sigmoid activation function was used for the last convolutional layer, which returned a 512×512 map as the binary mask of the detected diseased tissue.

To identify the salient image regions and amplify their influence while suppressing the irrelevant and confusing information, a more focused use of the feature maps was enforced. This was done by using an attention-based approach implemented with the inclusion of *Attention Gates* in the downsampling path, as described in more detail below.

9.5.3 Attention gates: structure and functioning

An AG was put on each skip connection that passes the feature maps from a downsampling layer d^i to the corresponding upsampling one u^i (see Fig. 9.11). It was used to prune irrelevant and noisy activation values in the stack of feature maps (i.e., the *dark gray* ones in 9.11) that are concatenated with the feature maps obtained by upsampling those of the previous layer (i.e., the *light gray* ones in 9.11).

The functioning of the AGs is illustrated in Fig. 9.12.

Considering the layer u^i, the corresponding AG takes in input the features maps of the previous layer u^{i+1} (with $u^{i+1} = p$ if $i = 3$) and those from the corresponding d^i. These stacks of features are firstly convoluted with a 1×1 kernel, to shrink all the maps into a stack with a fixed size Nf, with Nf computed as a quarter of the number of feature maps of u^{i+1}, so as to match, after the concatenation, the number of feature maps of u^{i+1}. Then, they are concatenated and passed through a ReLU activation layer and convoluted again with a kernel $1 \times 1 \times 1$ to obtain a single mask containing the attention coefficients for each pixel. After that, they are passed through a sigmoid activation layer. The resulting mask was used to multiply element-wise the feature maps from d^i.

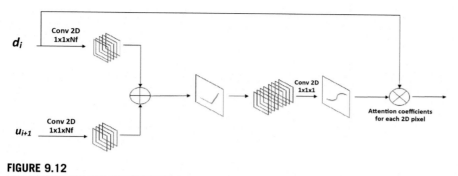

FIGURE 9.12

A schematic representation of an AG functioning.

9.5.4 Training details

The binary cross-entropy was selected as loss function, and hyperparameters were optimized with Adam optimizer, with a learning rate equal to 0.001, the exponential decay rates for the moving average of the gradient equal to 0.9, and the squared gradient equal to 0.999.

The Dice (D) score was used as performance metric, thus comparing the mask predicted by the Attention-FCN (P) with the ground-truth mask (G), as reported in Section 9.2.2, Eq. (9.2). The training runs on Keras (version 2.3.1) and TensorFlow frameworks (version 1.14.0) and was coded in Python 3.6. All experiments were performed under a Windows 10 OS on a machine with CPU Intel(R) Core(TM) i7−10700F CPU @ 2.90 GHz, GPU NVIDIA GeForce GTX 1650, and 32 GB of RAM.

The training set included a total of 17,177 sequential slices taken from 36 different HRCT scans. No data augmentation was applied to the images so the input samples were the original gray-level CT slices.

Considering the limited computational resources, we set the number of epochs to 50, and we saved the trained model at each epoch to test their performance afterward. It took about 550 s to complete just 1 epoch, so the entire training (50 epochs) took about 7 hours. See Fig. 9.13 for the loss and performance trends. The maximum value of the Dice score during training was 90.76% on the training set, and the minimum value of the loss function was 1.73%, both obtained at the 50th epoch.

9.5.5 Results

A test set of the remaining 7836 images (coming from the remaining 20 cases) was used to check the performance of the trained models. The highest Dice score value (computed as the mean score across all the test images) was 85.20%, and it was guaranteed by the model trained until the 50th epoch. All the other models trained for less epochs provided a lower Dice score.

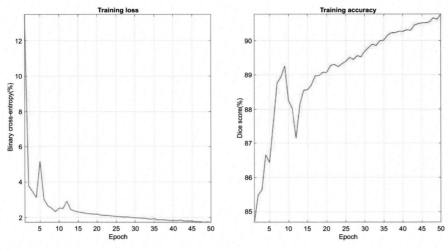

FIGURE 9.13

The binary cross-entropy function (left) and Dice score (right) monitored during the training.

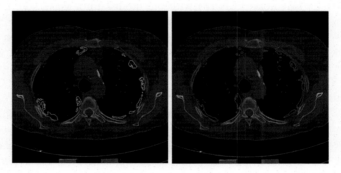

FIGURE 9.14

Left: the groud truth. Right: the segmentation of the Attention-FCNN.

Fig. 9.14 shows an example of the comparison between the ground truth, in green, and the segmentation produced by the Attention-FCNN model, in red. In order to assess the contribution of attention-based learning, we also run an ablation study, whose details are reported in the next section.

9.5.6 Ablation study

The ablation study aimed to investigate how the newly added AGs affect the segmentation performance.

Table 9.2 Dice score on training and test set of ablation study. Values in bold refer to the first two highest values obtained during the ablation study, both on the training set and the test set. They were highlighted with the aim of calculating the increase in performance obtained by putting an AG on each stage of the decoding path.

Stage with AG	Training set (%)	Test set (%)
p and u_1 and u_2 (proposed model)	**90.76**	**85.20**
u_1 and u_2	83.09	**82.39**
p and u_1	84.11	82.04
p and u_2	83.79	81.46
p	**85.04**	70.23
u_2	82.94	69.43
None	85.00	59.97
u_1	84.88	56.85

For a fair experimental comparison, we conducted the ablation study in exactly the same experiment environment as our main experiments presented in Section 9.4.3. We applied the ablation of AGs at various levels, thus comparing seven variants of our model: Attention-FCNN with an AG only in p, with an AG only in u_1, with an AG only in u_2, with AGs in p and u_1, with AGs in u_1 and u_2, with AGs in p and u_2, and finally without any AGs.

Table 9.2 summarizes the results of the study. From there, one can observe that incorporating AGs in the decoder pathway contributed to improve the segmentation performance by about 6,73% and 3,41% in terms of Dice score on the training and test sets, respectively, in comparison with the best results of the ablation study, achieved by the network with AGs on p and on u_1 and u_2, respectively.

As concluding remarks of our experiment, we argue that though our Attention-FCNN model was trained for few epochs, it guaranteed very good preliminary results. The use of the AGs was effectively beneficial and represented an added value to our network. Future works will entail more extensive experimentation for the binary segmentation of chest CT images and will be further transferred to a multi-class segmentation, able to differentiate among GGO, consolidations, and healthy tissue. Nevertheless, in order to ensure a real uptake in the clinical practice of our model as well as the others presented in the literature, many still open challenges are to be tackled, as discussed in the next concluding section of the chapter. Indeed, our method was trained on a monocentric, well-curated, homogeneous, and quality dataset, which enabled us to nurture a good training process. Further tests need to be done to verify the robustness of the model on diverse, noisy, and heterogeneous CT scans.

9.6 Discussion and conclusions

COVID-19 has spread throughout the world so quickly. CT imaging has proven to be an important tool in observing the typical manifestations of the disease and in monitoring its progression for the purpose of an adequate treatment plan. Nonetheless, radiologists lack automatic tools to accurately quantify the severity of COVID-19, i.e., the percentage of infection in the whole lung.

Over the past months, AI techniques have been extensively booming and deployed for a faster, objective, quantitative analysis of COVID-19 CT images. AI-based image segmentation tools may sensibly reduce radiologists' workload as they may automate the delineation and the quantification of the lung tissue that was damaged by the disease.

In this chapter, we explored lung-region- and lung-lesion-oriented DL methods to segment CT images of COVID-19 patients. We also presented a preliminary study in which an Attention-FCNN-based DL model was used to segment the whole pathological tissue in COVID-19 CT images, thus providing binary masks. Our results are in line with the literature, demonstrating that U-Net-inspired architectures (which is the gold standard in different medical image segmentation tasks) are promising to segment lung and lung COVID-19-related lesion.

However, there are still some open and critical issues:

- Imaging data often have incomplete or inexact labels. This causes an issue for building a robust segmentation model, as it is well known that the quality of the input will be mirrored in the quality of the trained model (i.e., the so-called *garbage-in, garbage-out* effect). Nevertheless, collecting huge and homogeneous datasets of correctly annotated cases is often a tough task for many reasons, among which the difficulty of sharing multicentric data belonging to different hospitals, and the inhomogeneity of image labels characterized by noise accidentally introduced by doctors during the annotation process.
- Data scarcity is another relevant issue, which makes training complex DL models difficult. Several approaches have been adopted so far to counteract this issue, such as *transfer learning*, namely, the use of networks pretrained with larger datasets, which allow for training, with the target dataset, only a reduced number of layers (often, the last ones). *Data augmentation* approach consists in applying geometrical transformation, like rotation, flip, or cropping, to the target images in order to increase their number. Finally, the *patch-wise* training approach, which consists of dividing the input training image into multiple overlapping or random patches, increases the number of training sample. However, adopting a patch-wise approach introduces the need to balance the dataset, namely, to have a comparable number of patches of healthy and diseased tissue to avoid the trained network being biased. In addition, the lack of contextual information deriving from the previous patches makes such an approach the least adopted.

- The parameters with which the images are reconstructed strongly depend on the acquisition device. This is reflected in different characteristics shown by the images that may bias and affect the performance of the model. For this reason, it would be preferable to have a training and validation set composed of a wide variety of samples. Nevertheless, if on the one hand the validation on multiple datasets may further improve the robustness and the generalization capabilities of the developed models, on the other hand, as mentioned above, it is not always possible to share images from different hospitals, due to both hardware limitations and privacy policies.

The implementation and the use of publicly, multicenter, available datasets may be a valuable strategy to thoroughly study the majority of the above-reported issues with the aim of solving them (see Section 3). The use of a larger amount of imaging data, acquired by different hospitals, would allow to obtain more robust segmentation models with higher generalization capability.

Moreover, the more reliable and robust the results are, the more confident clinicians will be in using DL-based segmentation models as a guideline in the diagnostic process. The question of the confidence of the medical community in the results deriving from DL methods provides the cue to introduce another future challenge. The growing need for interpretable models has paved the way for another branch of AI, called Explainable AI (XAI), which aims to provide an exhaustive explanation of the decision process followed by the predictive model by enhancing the image features that led to the prediction or segmentation. Given the severity of the impact of the pandemic and the urgent need for practical support from DL models, the development of an explainable framework has been temporarily inhibited. Certainly, in the future it will be of interest to provide, together with the segmentation results, also the explainable reasoning through which a DL model obtains the results.

References

[1] S. Manna, J. Wruble, S. Maron, D. Toussie, N. Voutsinas, M. Finkelstein, et al., COVID-19: a multimodality review of radiologic techniques, clinical utility, and imaging features, Radiol. Cardiothor. Imaging 2 (3) (2020). Available from: https://doi.org/10.1148/ryct.2020200210.

[2] H. Wong, H. Lam, A. Fong, S. Leung, T. Chin, C. Lo, et al., Frequency and distribution of chest radiographic findings in COVID-19 positive patients, Radiology 296 (2) (2019). Available from: https://doi.org/10.1148/radiol.2020201160.

[3] H. Koo, S. Lim, J. Choe, S. Choi, H. Sung, K. Do, Radiographic and CT features of viral pneumonia, Radiographics 38 (3) (2018) 719–739. Available from: https://doi.org/10.1148/rg.2018170048.

[4] J.P. Kanne, H. Bai, A. Bernheim, M. Chung, L.B. Haramati, D.F. Kallmes, et al., COVID-19 imaging: what we know now and what remains unknown, Radiology 299 (3) (2021) E262–E279. Available from: https://doi.org/10.1148/radiol.2021204522.

[5] J. Li, R. Yan, Y. Zhai, X. Qi, J. Lei, Chest CT findings in patients with coronavirus disease 2019 (COVID-19): a comprehensive review, Diagn. Interv. Radiol. 27 (5) (2021) 621−632. Available from: https://doi.org/10.5152/dir.2020.20212.

[6] X. Han, Y. Fan, O. Alwalid, N. Li, X. Jia, M. Yuan, et al., Six-month follow-up chest CT findings after severe COVID-19 pneumonia, Radiology 299 (1) (2021) E177−E186. Available from: https://doi.org/10.1148/radiol.2021203153.

[7] D. Dong, Z. Tang, S. Wang, H. Hui, L. Gong, Y. Lu, et al., The role of imaging in the detection and management of COVID-19: a review, IEEE. Rev. Biomed. Eng. 14 (2021) 16−29. Available from: https://doi.org/10.1109/RBME.2020.2990959.

[8] C. Huang, Y. Wang, X. Li, K. Lofy, J. Wiesman, H. Bruce, Clinical features of patients infected with 2019 novel coronavirus in wuhan, China Lancet (2020) 395.

[9] J. Kanne, B. Little, J. Chung, B. Elicker, L. Ketai, Outbreak of pneumonia of unknown etiology in wuhan china: the mystery and the miracle, Radiology 296 (2) (2020) 401−402. Available from: https://doi.org/10.1148/radiol.2020200527.

[10] M.E. Laino, A. Ammirabile, A. Posa, P. Cancian, S. Shalaby, V. Savevski, et al., The applications of artificial intelligence in chest imaging of COVID-19 patients: a literature review, Diagnostics 11 (8) (2021). Available from: https://doi.org/10.3390/diagnostics11081317.

[11] C. Kriza, V. Amenta, A. Zenié, D. Panidis, H. Chassaigne, P. Urbin, et al., Artificial intelligence for imaging-based COVID-19 detection: systematic review comparing added value of ai versus human readers, Eur. J. Radiol. 145 (2021) 110028. Available from: https://doi.org/10.1016/j.ejrad.2021.110028.

[12] H. Deng, X. Li, AI-empowered computational examination of chest imaging for COVID-19 treatment: a review, Front. Artif. Intelligence 4 (2021). Available from: https://doi.org/10.3389/frai.2021.612914.

[13] F. Shi, J. Wang, J. Shi, Z. Wu, Q. Wang, Z. Tang, et al., Review of artificial intelligence techniques in imaging data acquisition, segmentation, and diagnosis for covid-19, IEEE. Rev. Biomed. Eng. 14 (2021) 4−15. Available from: https://doi.org/10.1109/RBME.2020.2987975.

[14] J. Wang, X. Yang, B. Zhou, J.J. Sohn, J. Zhou, J.T. Jacob, et al., Review of machine learning in lung ultrasound in COVID-19 pandemic, J. Imaging 8 (3) (2022). Available from: https://doi.org/10.3390/jimaging8030065.

[15] A.B. Rosenkrantz, M. Mendiratta-Lala, B.J. Bartholmai, D. Ganeshan, R.G. Abramson, K.R. Burton, et al., Clinical utility of quantitative imaging, Acad. Radiol. 22 (1) (2015). Available from: https://doi.org/10.1016/j.acra.2014.08.011.

[16] E. Neri, V. Miele, F. Coppola, R. Grassi, Use of CT and artificial intelligence in suspected or COVID-19 positive patients: statement of the italian society, Radiol. Med. (Torino) 125 (5) (2020). Available from: https://doi.org/10.1007/s11547-020-01197-9.

[17] S. Inui, W. Gonoi, R. Kurokawa, Y. Nakai, Y. Watanabe, K. Sakurai, et al., The role of chest imaging in the diagnosis, management, and monitoring of coronavirus disease 2019 (COVID-19), Insights Imaging 12 (2021). Available from: https://doi.org/10.1186/s13244-021-01096-1.

[18] G. Rubin, C. Ryerson, L. Haramati, N. Sverzellati, J. Kanne, S. Raoof, et al., The role of chest imaging in patient management during the COVID-19 pandemic: a multinational consensus statement from the fleischner society, Chest 158 (2020) 106−116. Available from: https://doi.org/10.1016/j.chest.2020.04.003.

[19] H. Koo, S. Lim, J. Choe, S. Choi, H. Sung, K. Do, Essentials for radiologists on COVID-19: an update—radiology scientific expert panel, J. Med. Virol. 92 (4) (2020) 401−402.

[20] M. Carotti, F. Salaffi, P. Sarzi-Puttini, A. Agostini, A. Borgheresi, D. Minorati, Chest CT features of coronavirus disease 2019 (COVID-19) pneumonia: key points for radiologists, Radiol. Med. 125 (7) (2020) 636−646.

[21] W. Yang, A. Sirajuddin, X. Zhang, G. Liu, Z. Teng, S. Zhao, et al., The role of imaging in 2019 novel coronavirus pneumonia (covid-19), Eur. Radiol. 4 (1) (2020). Available from: https://doi.org/10.1007/s00330-020-06827-4.

[22] Z. Ye, Y. Zhang, Y. Wang, Z. Huang, B. Song, Chest CT manifestations of new coronavirus disease 2019 (covid-19): a pictorial review, Eur. Radiol. 23 (1) (2020). Available from: https://doi.org/10.1007/s00330-020-06801-0.

[23] V.K. Venugopal, V. Mahajan, S. Rajan, V. Agarwal, R. Rajan, S. Syed, et al., A Systematic Meta-Analysis of CT Features of COVID-19: Lessons from Radiology. Available online at: https://doi.org/10.1101/2020040420052241; 2020.

[24] A. El-Sherief, M. Gilman, T. Healey, R. Tambouret, J. Shepard, G. Abbott, et al., Clear vision through the haze: a practical approach to ground-glass opacity, Curr. Probl. Diagn. Radiol. 43 (3) (2014) 140−158. Available from: https://doi.org/10.1067/j.cpradiol.2014.01.004.

[25] P. Chatterjee, D.S. Rani, A survey on techniques used in medical imaging processing, J. Phys. Conf. Ser. 2089 (1) (2021) 012013. Available from: https://doi.org/10.1088/1742-6596/2089/1/012013.

[26] D.L. Pham, C. Xu, J.L. Prince, Current methods in medical image segmentation, Annu. Rev. Biomed. Eng. 2 (1) (2000) 315−337. Available from: https://doi.org/10.1146/annurev.bioeng.2.1.315.

[27] N. Otsu, A threshold selection method from gray-level histograms, IEEE. Trans. Syst. Man. Cybern. 9 (1) (1979) 62−66. Available from: https://doi.org/10.1109/TSMC.1979.4310076.

[28] R.C. Gonzale, R.E. Woods, Digital Image Processing, fourth ed., Pearson, 2018.

[29] Z. Yu-qian, G. Wei-hua, C. Zhen-cheng, T. Jing-tian, L. Ling-yun, Medical images edge detection based on mathematical morphology, in: 2005 IEEE Engineering in Medicine and Biology 27th Annual Conference, 2005, pp. 6492−6495. Available from: https://doi.org/10.1109/IEMBS.2005.1615986.

[30] S. Pohlman, K.A. Powell, N.A. Obuchowski, W.A. Chilcote, S. Grundfest-Broniatowski, Quantitative classification of breast tumors in digitized mammograms, Med. Phys. 23 (8) (1997). Available from: https://doi.org/10.1118/1.597707.

[31] S. Huang, S.H. Ong, K.C. Foong, P.S. Goh, W.L. Nowinski, Medical image segmentation using watershed segmentation with texture-based region merging, in: Proceedings of the Annual International Conference of the IEEE Engineering in Medicine and Biology Society. IEEE Engineering in Medicine and Biology Society. Annual International Conference, no., 2008, pp. 4039−42. Available from: https://doi.org/10.1109/IEMBS.2008.4650096.

[32] R. Malladi, J. Sethian, B. Vemuri, Shape modeling with front propagation: a level set approach, IEEE. Trans. Pattern. Anal. Mach. Intell. 17 (2) (1995) 158−175. Available from: https://doi.org/10.1109/34.368173.

[33] Y. Ebrahimdoost, J. Dehmeshki, T.S. Ellis, M. Firoozbakht, A. Youannic, S. Qanadli, Medical image segmentation using active contours and a level set

model: application to pulmonary embolism (pe) segmentation, in: 2010 Fourth International Conference on Digital Society, 2010, pp. 269−273. Available from: https://doi.org/10.1109/ICDS.2010.64.

[34] P. Swierczynski, B.W. Papież, J.A. Schnabel, C. Macdonald, A level-set approach to joint image segmentation and registration with application to ct lung imaging, Comput. Med. Imaging Graph. 65 (2018) 58−68. Available from: https://doi.org/10.1016/j.compmedimag.2017.06.003.

[35] N.J. Tustison, K. Qing, C. Wang, T.A. Altes, J.Pr Mugler, Atlas-based estimation of lung and lobar anatomy in proton mri, Magn. Reson. Med. 76 (1) (2016) 315−320. Available from: https://doi.org/10.1002/mrm.25824.

[36] K. Held, E. Kops, B. Krause, W. Wells, R. Kikinis, H.W. Muller-Gartner, Markov random field segmentation of brain mr images, IEEE Trans. Med. Imaging 16 (6) (1997) 878−886. Available from: https://doi.org/10.1109/42.650883.

[37] Y. Tan, L.H. Schwartz, B. Zhao, Segmentation of lung lesions on ct scans using watershed, active contours, and markov random field, Med. Phys. 40 (4) (2013). Available from: https://doi.org/10.1118/1.4793409.

[38] S. Colantonio, O. Salvetti, I.B. Gurevich, A two-step approach for automatic microscopic image segmentation using fuzzy clustering and neural discrimination, Pattern Recog. Image Anal. 17 (3) (2007). Available from: https://doi.org/10.1134/S1054661807030108.

[39] D. Moroni, S. Colantonio, O. Salvetti, M. Salvetti, Heart deformation pattern analysis through shape modelling, Pattern Recog. Image Anal. 19 (2) (2009). Available from: https://doi.org/10.1134/S1054661809020084.

[40] D. Pham, J. Prince, An adaptive fuzzy c-means algorithm for image segmentation in the presence of intensity inhomogeneities, Pattern Recog. Lett. (1999) 57−69.

[41] S. Colantonio, D. Moroni, O. Salvetti, Mri left ventricle segmentation and reconstruction for the study of the heart dynamics, in: Proceedings of the Fifth IEEE International Symposium on Signal Processing and Information Technology, 2005, pp. 213−218. Available from: https://doi.org/10.1109/ISSPIT.2005.1577098.

[42] S. Colantonio, I.B. Gurevich, O. Salvetti, Automatic fuzzy-neural based segmentation of microscopic cell images, Int. J. Signal Imaging Syst. Eng. 1 (1) (2008) 18−24.

[43] S.N. Kumar, A. Lenin Fred, P.S. Varghese, An overview of segmentation algorithms for the analysis of anomalies on medical images, J. Intelligent Syst. 29 (1) (2020) 612−625. Available from: https://doi.org/10.1515/jisys-2017-0629.

[44] Y. LeCun, Y. Bengio, G. Hinton, Deep learning, Nature 521 (2015) 436−444.

[45] E. Bisong, Regularization for Deep Learning. Apress. ISBN 978-1-4842-4470-8, 2019; pp. 415−421. Available from: https://doi.org/10.1007/978-1-4842-4470-8_34.

[46] S. Colantonio, A. Salvati, C. Caudai, F. Bonino, L.D. Rosa, M.A. Pascali, et al., A deep learning approach for hepatic steatosis estimation from ultrasound imaging, in: K. Wojtkiewicz, J. Treur, E. Pimenidis, M. Maleszka (Eds.), Advances in Computational Collective Intelligence, Springer International Publishing, Cham, 2021, pp. 703−714.

[47] A. Prasoon, K. Petersen, C. Igel, F. Lauze, E. Dam, M. Nielsen, Deep feature learning for knee cartilage segmentation using a triplanar convolutional neural network, in: K. Mori, I. Sakuma, Y. Sato, C. Barillot, N. Navab (Eds.), Medical Image Computing and Computer-Assisted Intervention, MICCAI, 2013, pp. 246−253.

[48] M. Lyksborg, O. Puonti, M. Agn, R. Larsen, An Ensemble of 2d Convolutional Neural Networks for Tumor Segmentation (R.R. Paulsen, K.S. Pedersen, Eds.), Image Analysis, 2015, pp. 201−211.

[49] R. Wang, T. Lei, R. Cui, B. Zhang, H. Meng, A.K. Nandi, Medical image segmentation using deep learning: a survey, IET Image Process. 16 (5) (2022) 1243–1267. Available from: https://doi.org/10.1049/ipr2.12419.

[50] C. Chakraborty, P. Malhotra, S. Gupta, D. Koundal, A. Zaguia, W. Enbeyle, Deep neural networks for medical image segmentation, J. Healthc. Eng. (2022). Available from: https://doi.org/10.1155/2022/9580991.

[51] J. Long, E. Shelhamer, T. Darrell, Fully convolutional networks for semantic segmentation, in: 2015 IEEE Conference on Computer Vision and Pattern Recognition (CVPR), 2015, pp. 3431–3440. Available from: https://doi.org/10.1109/CVPR.2015.7298965.

[52] K. Kamnitsas, C. Ledig, V.F. Newcombe, J.P. Simpson, A.D. Kane, D.K. Menon, et al., Efficient multi-scale 3D CNN with fully connected CRF for accurate brain lesion segmentation, Med. Image. Anal. 36 (2017) 61–78. Available from: https://doi.org/10.1016/j.media.2016.10.004.

[53] P. Krähenbühl, V. Koltun, Efficient inference in fully connected crfs with gaussian edge potentials, in: J. Shawe-Taylor, R. Zemel, P. Bartlett, F. Pereira, K. Weinberger (Eds.), Advances in Neural Information Processing Systems, vol. 24, 2011.

[54] O. Ronneberger, P. Fischer, T. Brox, U-net: convolutional networks for biomedical image segmentation, in: N. Navab, J. Hornegger, W.M. Wells, A.F. Frangi (Eds.), Medical Image Computing and Computer-Assisted Intervention — MICCAI, Springer International Publishing, 2015.

[55] Z. Zhou, M.M.R. Siddiquee, N. Tajbakhsh, J. Liang, Unet++: redesigning skip connections to exploit multiscale features in image segmentation, IEEE Trans. Med. Imaging 39 (6) (2020). Available from: https://doi.org/10.1109/TMI.2019.2959609.

[56] C. Guo, M. Szemenyei, Y. Pei, Y. Yi, W. Zhou, Sd-unet: a structured dropout u-net for retinal vessel segmentation, in: 2019 IEEE 19th International Conference on Bioinformatics and Bioengineering (BIBE), 2019, pp. 439–444. Available from: https://doi.org/10.1109/BIBE.2019.00085.

[57] Q. Jin, Z. Meng, C. Sun, H. Cui, R. Su, Ra-unet: a hybrid deep attention-aware network to extract liver and tumor in ct scans, Front. Bioeng. Biotechnol. 8 (2020). Available from: https://www.frontiersin.org/article/10.3389/fbioe.2020.605132.

[58] F. Isensee, P.F. Jaeger, S.A. Kohl, J. Petersen, K.H. Maier-Hein, Nnu-net: a self-configuring method for deep learning-based biomedical image segmentation, Nat. Methods 18 (2) (2021). Available from: https://doi.org/10.1038/s41592-020-01008-z.

[59] H. Roth, Z. Xu, C.T. Diez, R.S. Jacob, J. Zember, J. Molto, et al., Rapid artificial intelligence solutions in a pandemic - the covid-19–20 lung ct lesion segmentation challenge, Res. Square (2021). Available from: https://doi.org/10.21203/rs.3.rs-571332/v1.

[60] L.C. Chen, G. Papandreou, I. Kokkinos, K. Murphy, A.L. Yuille, Deeplab: semantic image segmentation with deep convolutional nets, atrous convolution, and fully connected crfs, 2016. Available from: https://doi.org/10.48550/ARXIV.1606.00915.

[61] L.C. Chen, G. Papandreou, I. Kokkinos, K. Murphy, A.L. Yuille, Deeplab: semantic image segmentation with deep convolutional nets, atrous convolution, and fully connected crfs, IEEE. Trans. Pattern. Anal. Mach. Intell. 40 (4) (2018) 834–848. Available from: https://doi.org/10.1109/TPAMI.2017.2699184.

[62] L.C. Chen, G. Papandreou, F. Schroff, H. Adam, Rethinking atrous convolution for semantic image segmentation, 2017. Available from: https://doi.org/10.48550/ARXIV.1706.05587, https://arxiv.org/abs/1706.05587.

[63] L.C. Chen, Y. Zhu, G. Papandreou, F. Schroff, H. Adam, Encoder-decoder with atrous separable convolution for semantic image segmentation, in: Proceedings of the European Conference on Computer Vision (ECCV), 2018.

[64] J. Wang, X. Liu, Medical image recognition and segmentation of pathological slices of gastric cancer based on deeplab v3 + neural network, Comput. Methods Programs Biomed. 207 (2021) 106210. Available from: https://doi.org/10.1016/j.cmpb.2021.106210.

[65] Y. Yue, Z. Tian, Y. Qiao, Transdeeplabv3: multi-prior segmentor for medical image segmentation, in: 2021 China Automation Congress (CAC), 2021, pp. 6880–6885. Available from: https://doi.org/10.1109/CAC53003.2021.9727997.

[66] R. Girshick, J. Donahue, T. Darrell, J. Malik, Rich feature hierarchies for accurate object detection and semantic segmentation, in: 2014 IEEE Conference on Computer Vision and Pattern Recognition, 2014, pp. 580–587. Available from: https://doi.org/10.1109/CVPR.2014.81.

[67] R. Girshick, Fast r-cnn, 2015. Available from: https://doi.org/10.48550/ARXIV.1504.08083, https://arxiv.org/abs/1504.08083.

[68] S. Ren, K. He, R. Girshick, J. Sun, Faster r-cnn: towards real-time object detection with region proposal networks, in: C. Cortes, N. Lawrence, D. Lee, M. Sugiyama, R. Garnett (Eds.), Advances in Neural Information Processing Systems, vol. 28, Curran Associates, Inc., 2015.

[69] K.H. Shibly, S.K. Dey, T.U. Md Islam, M. Md Rahman, Covid faster r-cnn: a novel framework to diagnose novel coronavirus disease (covid-19) in X-ray images, Inform. Med. Unlocked 20 (2020) 100405. Available from: https://doi.org/10.1016/j.imu.2020.100405.

[70] J. Long, E. Shelhamer, T. Darrell, Fully convolutional networks for semantic segmentation. arXiv 2014; Available from: https://doi.org/10.48550/ARXIV.1411.4038.

[71] A. Taha, A. Hanbury, Metrics for evaluating 3d medical image segmentation: analysis, selection, and tool, BMC Med. Imaging 15 (2015). Available from: https://doi.org/10.1186/s12880-015-0068-x.

[72] R. Marie-Pierre, P.P. Anagha, P. Helmut, S. Mario, S. Nicola, G. Fergus, et al., COVID-19 patients and the radiology department - advice from the european society of radiology (esr) and the european society of thoracic imaging (esti), Eur. Radiol. 30 (9) (2020) 4903–4909. Available from: https://doi.org/10.1007/s00330-020-06865-y.

[73] MosMedData: chest CT scans with COVID-19 related findings, 2020. Available from: https://mosmed.ai/en/.

[74] Covid-19 ct segmentation dataset, 2020. Available from: https://medicalsegmentation.com/covid19/.

[75] M. Jun, G. Cheng, W. Yixin, A. Xingle, G. Jiantao, Y. Ziqi, et al., COVID-19 CT Lung and Infection Segmentation Dataset, 2020. Available from: https://doi.org/10.5281/zenodo.3757476, https://doi.org/10.5281/zenodo.3757476.

[76] M. de la Iglesia Vayá, J.M. Saborit-Torres, J.A. Montell Serrano, E. Oliver-Garcia, A. Pertusa, A. Bustos, et al., Bimcv COVID-19 + : a large annotated dataset of rx and CT images from COVID-19 patients, 2021. Available from: https://doi.org/10.21227/w3aw-rv39.

[77] L. Maiello, L. Ball, M. Micali, F. Iannuzzi, N. Scherf, R.T. Hoffmann, et al., Automatic lung segmentation and quantification of aeration in computed tomography of the chest using 3d transfer learning, Front. Physiol. 12 (2022). Available from: https://doi.org/10.3389/fphys.2021.725865.

[78] S. Tilborghs, I. Dirks, L. Fidon, S. Willems, T. Eelbode, J. Bertels, et al., Comparative study of deep learning methods for the automatic segmentation of lung, lesion and lesion type in ct scans of covid-19 patients, 2020. Available from: https://doi.org/10.48550/ARXIV.2007.15546.

[79] A.A.A. Setio, A. Traverso, T. de Bel, M.S.N. Berens, C. van den Bogaard, P. Cerello, et al., Validation, comparison, and combination of algorithms for automatic detection of pulmonary nodules in computed tomography images: the luna16 challenge, Med. Image. Anal. 42 (2017) 1−13.

[80] W. Xie, C. Jacobs, J.P. Charbonnier, B. van Ginneken, Relational modeling for robust and efficient pulmonary lobe segmentation in CT scans, IEEE Trans. Med. Imaging 39 (8) (2020) 2664−2675. Available from: https://doi.org/10.1109/tmi.2020. 2995108.

[81] Y. Wang, Y. Zhang, Y. Liu, J. Tian, C. Zhong, Z. Shi, et al., Does non-COVID-19 lung lesion help? investigating transferability in COVID-19 CT image segmentation, Comput. Methods Programs Biomed. 202 (2021) 106004. Available from: https://doi.org/10.1016/j.cmpb.2021.106004.

[82] S. Napel, S.K. Plevritis, Nsclc radiogenomics: initial stanford study of 26 cases, 2014. Available from: https://doi.org/10.7937/K9/TCIA.2014.X7ONY6B1.

[83] S. Bakr, O. Gevaert, Echegaray Sea, A radiogenomic dataset of non-small cell lung cancer, Sci. Data 5 (2018). Available from: https://doi.org/10.1038/sdata.2018.202.

[84] O. Gevaert, J. Xu, C.D. Hoang, A.N. Leung, Y. Xu, A. Quon, et al., Non-small cell lung cancer: identifying prognostic imaging biomarkers by leveraging public gene expression microarray data−methods and preliminary results, Radiology 264 (2) (2012) 387−396. Available from: https://europepmc.org/articles/PMC3401348.

[85] X. Li, W. Wang, X. Hu, J. Yang, Selective kernel networks, 2019. Available from: https://doi.org/10.48550/ARXIV.1903.06586.

[86] L. Jiannan, D. Bo, W. Shuai, C. Hui, F. Deng-Ping, M. Jiquan, et al., COVID-19 lung infection segmentation with a novel two-stage cross-domain transfer learning framework, Med. Image. Anal. 74 (2021) 102205. Available from: https://doi.org/10.1016/j.media.2021.102205.

[87] J. Deng, W. Dong, R. Socher, L.J. Li, K. Li, L. Fei-Fei, Imagenet: a large-scale hierarchical image database, in: 2009 IEEE Conference on Computer Vision and Pattern Recognition, 2009, pp. 248−255. Available from: https://doi.org/10.1109/CVPR. 2009.5206848.

[88] S. Armato, G. McLennan, L. Bidaut, M. McNitt-Gray, C. Meyer, A. Reeves, et al., The lung image database consortium (lidc) and image database resource initiative (idri): a completed reference database of lung nodules on CT scans, Med. Phys. 38 (2) (2011).

[89] C. Guillaume, V. Maria, B. Enzo, C. Stergios, H.T. Trieu-Nghi, D. Severine, et al., AI-driven quantification, staging and outcome prediction of COVID-19 pneumonia, Med. Image. Anal. 67 (2021) 101860. Available from: https://doi.org/10.1016/j.media.2020.101860.

[90] M. Vakalopoulou, G. Chassagnon, N. Bus, R.M. Silva, E. Zacharaki, M.P. Revel, et al., Atlasnet: multi-atlas non-linear deep networks for medical image segmentation, in: MICCAI, 2018.

[91] Ö. Çiçek, A. Abdulkadir, S.S. Lienkamp, T. Brox, O. Ronneberger, 3d u-net: learning dense volumetric segmentation from sparse annotation, in: S. Ourselin, L. Joskowicz, M.R. Sabuncu, G. Unal, W. Wells (Eds.), Medical Image Computing and Computer-Assisted Intervention, MICCAI, 2016.

[92] V. Badrinarayanan, A. Kendall, R. Cipolla, Segnet: a deep convolutional encoder-decoder architecture for image segmentation, IEEE. Trans. Pattern. Anal. Mach. Intell. 39 (12) (2017) 2481−2495. Available from: https://doi.org/10.1109/TPAMI. 2016.2644615.

[93] S. Chaganti, P. Grenier, A. Balachandran, G. Chabin, S. Cohen, T. Flohr, et al., Automated quantification of CT patterns associated with COVID-19 from chest CT, Radiol. Artif. Intelligence 2 (4) (2020) e200048. Available from: https://doi.org/ 10.1148/ryai.2020200048.

[94] Y. Cao, S. Liu, Y. Peng, J. Li, Denseunet: densely connected unet for electron microscopy image segmentation, IET Image Process. 14 (12) (2020) 2682−2689. Available from: https://doi.org/10.1049/iet-ipr.2019.1527.

[95] Ct images in covid-19 - the cancer imaging archive (tcia) public access - cancer imaging archive wiki - accessed april 2022. Available from: https://doi.org/10.7937/ tcia.2020.gqry-nc81.

[96] Chest imaging with clinical and genomic correlates representing a rural covid-19 positive population (covid-19-ar) - the cancer imaging archive (tcia) public access - cancer imaging archive wiki - accessed april 2022. Available from: https://doi.org/ 10.7937/tcia.2020.gqry-nc81.

[97] Nvidia ngc. Accessed April 2022. Available from: https://ngc.nvidia.com/catalog/ models/nvidia:clara_train_covid19_ct_lesion_seg.

[98] D.P. Fan, T. Zhou, G.P. Ji, Y. Zhou, G. Chen, H. Fu, et al., Inf-net: automatic COVID-19 lung infection segmentation from CT images, IEEE Trans. Med. Imaging 39 (8) (2020) 2626−2637. Available from: https://doi.org/10.1109/TMI. 2020.2996645.

[99] J. Long, E. Shelhamer, T. Darrell, Fully convolutional networks for semantic segmentation, in: 2015 IEEE Conference on Computer Vision and Pattern Recognition (CVPR), IEEE Computer Society, 2015. Available from: https://doi.org/10.1109/ CVPR.2015.7298965.

[100] J.P. Cohen, P. Morrison, L. Dao, Covid-19 image data collection, 2020. Available from: https://doi.org/10.48550/ARXIV.2003.11597.

[101] R. Buongiorno, D. Germanese, C. Romei, L. Tavanti, A. Liperi, S. Colantonio, UIP-Net: a decoder-encoder CNN for the detection and quantification of usual interstitial pneumoniae pattern in lung CT scan images, 2021; pp. 389−405. Available from: https://doi.org/10.1007/978-3-030-68763-2_30.

Index

Note: Page numbers followed by "*f*" and "*t*" refer to figures and tables, respectively.